PROBLEMS IN ORGANIC CHEMISTRY

PROBLEMS IN ORGANIC CHEMISTRY
A Lead-Oriented Approach

DAVID E. HORN
Goucher College

MICHAEL J. STRAUSS
University of Vermont

JOHN WILEY & SONS
New York • Chichester • Brisbane • Toronto • Singapore

Copyright © 1986, by John Wiley & Sons, Inc.

All rights reserved. Published simultaneously in Canada.

Reproduction or translation of any part of
this work beyond that permitted by Sections
107 and 108 of the 1976 United States Copyright
Act without the permission of the copyright
owner is unlawful. Requests for permission
or further information should be addressed to
the Permissions Department, John Wiley & Sons.

Library of Congress Cataloging-in-Publication Data:

Horn, David E.
   Problems in organic chemistry.

   Includes index.
   1. Chemistry, Organic—Problems, exercises, etc.
I. Strauss, Michael J.   II. Title.
QD257.H67 1986   547.0076   85-12476
ISBN 0-471-81649-3 (pbk.)

Printed in the United States of America

10 9 8 7 6 5 4 3 2 1

To Amanda, Jennifer, Matthew, Merissa, Roberta, and Susan

PREFACE

This is a different kind of organic chemistry problem-solving book. It uses a lead-oriented incremental approach that allows students to solve organic chemistry problems in a manner that best facilitates learning. It complements the lecture material by sequentially building concepts within a problem-solving context. Specific areas of interest may be chosen for learning. When it is used correctly, the book approximates, in a simple fashion, a student–instructor dialogue.

Conventional problem-solving books allow students only three options. First, they can solve the problem. However, if they are unable to solve it, they have two other options. Either they can go to the textbook to try, in an unguided fashion, to find assistance in the form of analogy, or they can look up the answer in the back of the book. This last option is a particularly poor alternative. By providing too much information it destroys much of the learning that might have occurred. Often students can solve problems with appropriate leads or clues, similar to those given by an instructor in a problem review session, but with the complete solution in front of them this opportunity is lost. Looking up the answer destroys the connections that are being built during the working out of a solution.

If, however, students are guided to the correct answers by leads or hints, then the opportunities for learning increase. These leads postpone the sometimes irresistable temptation to look up the answer before all possibilities have been investigated. Our book uses this unique approach to solving problems.

Chapter 1 consists of some basic principles for solving problems in organic chemistry. These principles are scattered throughout leading organic texts, but they are included here in brief form for handy reference and review.

Chapter 2 consists of a collection of problems. The sequence of problems runs parallel to the order of topics in *Organic Chemistry and Fundamentals of Organic Chemistry* by T. W. Graham Solomons. In addition, to make the book compatible with other organic texts, the name of the pertinent chapter (or topic) is included with each problem. The degree of difficulty assigned to each problem is designated by the number of asterisks (*, least difficult; **, moderately difficult; ***, most difficult).

Chapter 3 consists of verbal leads (clues) for each problem found in Chapter 2. Chapter 4 contains detailed solutions with chemical structures and mechanistic schemes, as well as in-depth discussions where appropriate.

This book should be useful at all levels of *organic* chemistry. Premedical students reviewing for admission examinations and students in allied health fields should find it valuable. In addition, it is our hope that majors in chemistry, biology, pharmacology, and biochemistry will find this book an indispensable complement to any traditional *organic* text.

PREFACE

We thank Gilbert Grady of Saint Michael's College for his helpful comments and constructive criticism, David Evelti for his efforts on the structural drawings, and members of our respective chemistry departments for their support. Finally, we thank Susan Platt for typing the manuscript.

David E. Horn
Michael J. Strauss

CONTENTS

Chapter 1 BASIC PRINCIPLES FOR PROBLEM SOLVING
          IN ORGANIC CHEMISTRY   1

Chapter 2 PROBLEMS   17

Chapter 3 VERBAL LEADS   72

Chapter 4 SOLUTIONS WITH STRUCTURAL FORMULAS   107

Index to Problems   195

1 BASIC PRINCIPLES FOR PROBLEM SOLVING IN ORGANIC CHEMISTRY

There are three general types of problems in this book: structure determination, synthesis, and mechanism. These topics are not mutually exclusive, but they represent a traditional classification of problems found in *Organic Chemistry* by T.W. Graham Solomons. The purpose of this chapter is to provide a brief description of the fundamentals underlying successful solution of these problems.

STRUCTURE DETERMINATION

Determination of the structure of an organic compound requires a basic understanding of the implications of spectroscopic data. Extensive recall of infrared (IR) and NMR spectral data is not necessarily required to do these problems. A brief return to the spectroscopy section of the main text should provide the specific information needed. When an understanding of spectroscopic data is combined with knowledge of reaction mechanisms or, in some cases, associated with named reactions, further leads concerning the identity of the compounds in question should become apparent.

SYNTHESIS

Designing an organic synthesis can be accomplished by posing several pertinent questions and considering possible answers. Specific details, such as, which reactants and reagents are needed, can be found in the body of the main text. What follows is a list of general questions which, when answered correctly, lead to possible synthetic routes to an organic compound.

1. What starting materials are available for the synthesis?
2. Is a change in the carbon skeleton of the starting material necessary to yield the product?
    (a) If starting material containing the correct carbon skeleton is available, what changes in functionality are required to form the product? What reagents will accomplish this change?
    (b) If the starting material contains an insufficient number of carbon atoms, how

2 BASIC PRINCIPLES FOR PROBLEM SOLVING IN ORGANIC CHEMISTRY

can the chain be lengthened? What reactions will produce the proper carbon skeleton?

3. Will working backward simplify the choice of a starting material? (Dividing the carbon chain of the compound to be synthesized into smaller units, which are available in the starting material, often provides ideas for a potential synthetic route.)

The flow chart that follows summarizes these questions.

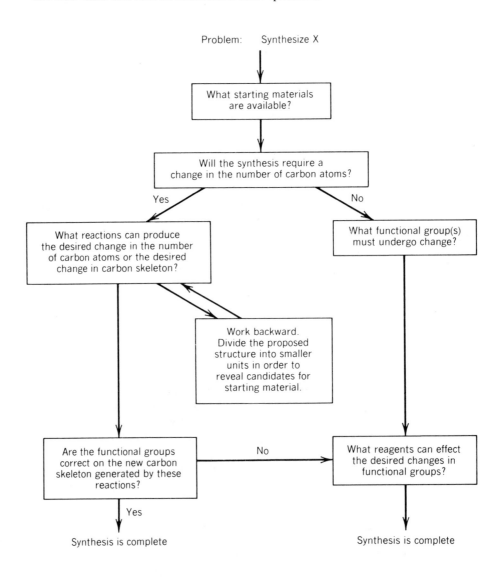

PRINCIPLES FOR DETERMINING REACTION MECHANISMS

Solving problems in organic reaction mechanisms requires familiarity with various types of organic reactions as well as knowledge of the different intermediates, which might be encountered along the reaction path. Specific fundamental principles must then be applied to support or refute the proposed mechanism or intermediate. The discussion that follows is a brief review of reaction types, intermediates, and fundamental principles needed for a successful approach to this type of problem.

Types of Organic Reactions

Students of organic chemistry soon realize that organic reactions can be classified as follows:

1. Substitution
2. Addition
3. Elimination
4. Rearrangement

Substitution

A typical substitution reaction involves the replacement of one group (or atom) in the starting material by another group (or atom). In the following reaction chlorine replaces hydrogen.

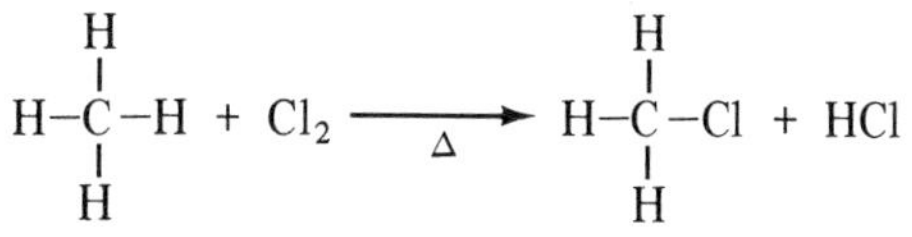

$$CH_4 + Cl_2 \xrightarrow{\Delta} CH_3Cl + HCl$$

Another example shows hydroxide substituting for iodide ion.

$$CH_3I + Na^{\oplus}OH^{\ominus} \longrightarrow CH_3OH + Na^{\oplus}I^{\ominus}$$

In yet another example, the nitro group replaces hydrogen.

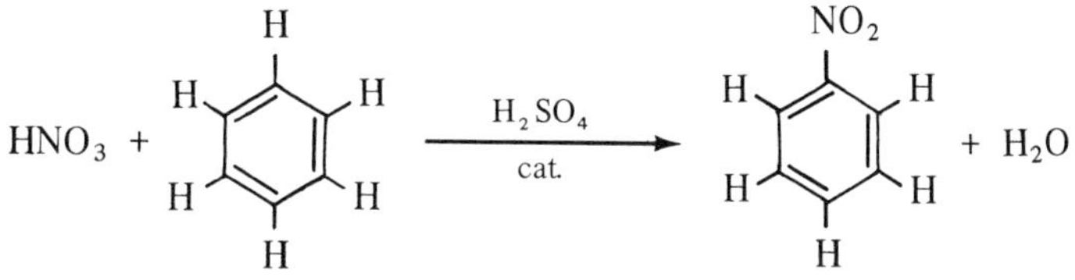

$$HNO_3 + C_6H_6 \xrightarrow[\text{cat.}]{H_2SO_4} C_6H_5NO_2 + H_2O$$

Addition

An addition reaction involves the interaction of an appropriate reagent with a multiple bond such as a carbon–carbon double bond, a carbon–oxygen double bond, or

4 BASIC PRINCIPLES FOR PROBLEM SOLVING IN ORGANIC CHEMISTRY

carbon–carbon triple bond, and so forth. In the following first example, two bromine atoms originating from Br_2 add to the pi system to produce two new sigma bonds (with concomitant loss of the double bond).

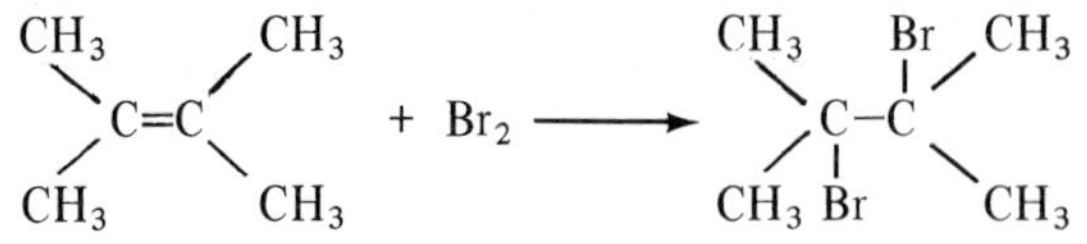

$$(CH_3)_2C=C(CH_3)_2 + Br_2 \longrightarrow (CH_3)_2CBr-CBr(CH_3)_2$$

The product is clearly an adduct (addition product) since nothing else is formed. Similarly, CH_3OH adds to CH_3-CHO to produce a product that contains both CH_3OH and CH_3-CHO. Therefore this also represents an addition reaction.

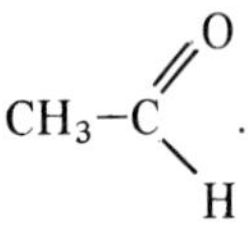

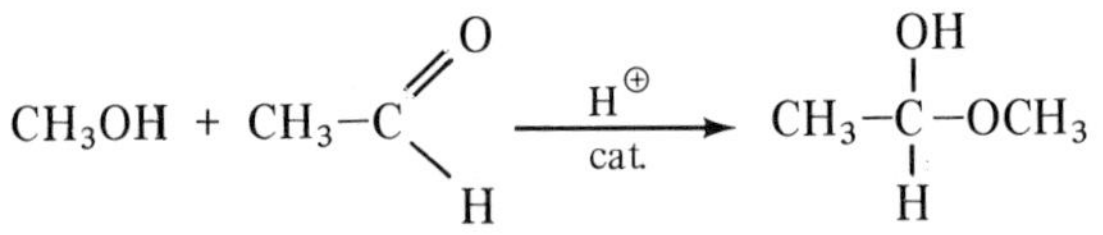

$$CH_3OH + CH_3-CHO \xrightarrow[\text{cat.}]{H^{\oplus}} CH_3-CH(OH)-OCH_3$$

Elimination

If a reaction proceeds to product by losing part of the reactant, it represents an elimination reaction. In the following reaction, the elements of HCl are lost (eliminated) from the starting material:

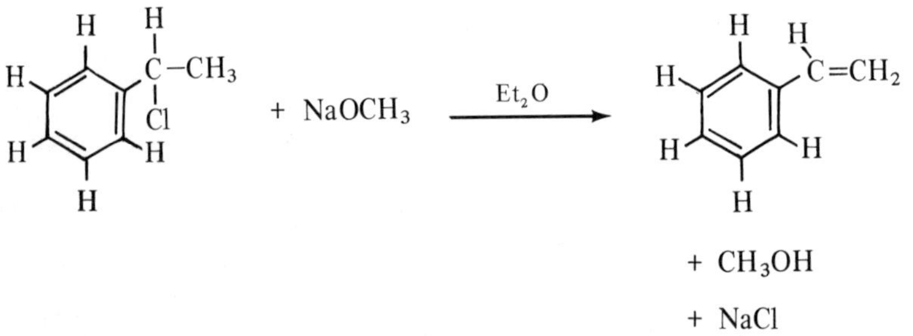

$$\text{C}_6\text{H}_5\text{-CHCl-CH}_3 + \text{NaOCH}_3 \xrightarrow{Et_2O} \text{C}_6\text{H}_5\text{-CH=CH}_2 + CH_3OH + NaCl$$

In another example of an elimination, water is lost simply by heating the reactant, an alcohol.

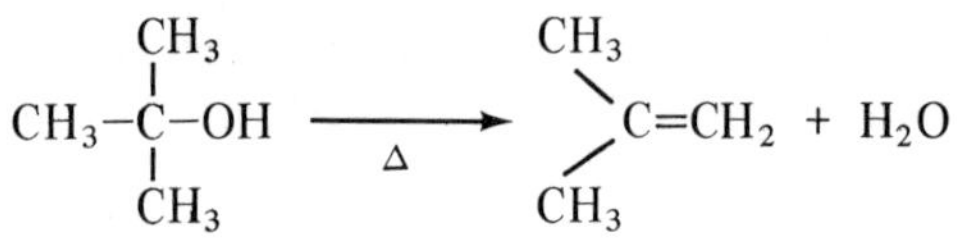

$$(CH_3)_3C-OH \xrightarrow{\Delta} (CH_3)_2C=CH_2 + H_2O$$

Rearrangement

In this final category of reaction types there are several interesting examples that all have a common theme, a change in carbon skeleton. In the first example, a carbon skeleton rearrangment occurs by expansion of a four- to a five-membered ring.

$$\begin{array}{c} CH_2 \\ | \\ CH_2 \end{array} \!\!\! \begin{array}{c} \diagdown \\ \diagup \end{array} \!\! C \!\! \begin{array}{c} \diagup CH_3 \\ \diagdown CH_2OH \end{array} \quad \xrightarrow[\text{cat.}]{H^{\oplus}} \quad \begin{array}{c} CH_2 \\ | \\ CH_2 \end{array} \!\!\! \begin{array}{c} \diagdown \\ \diagup \end{array} \!\! C \!\! \begin{array}{c} \diagup CH_3 \\ \diagdown OH \end{array}$$

In another example, a methyl group ($-CH_3$) in the starting diol migrates, and a ketone is the product.

$$\underset{\text{Diol}}{CH_3 - \underset{\underset{CH_3}{|}}{\overset{\overset{OH}{|}}{C}} - \underset{\underset{CH_3}{|}}{\overset{\overset{OH}{|}}{C}} - CH_3} \quad \xrightarrow[\text{cat.}]{H^{\oplus}} \quad \underset{\text{Ketone}}{CH_3 - \overset{\overset{O}{\|}}{C} - \underset{\underset{CH_3}{|}}{\overset{\overset{CH_3}{|}}{C}} - CH_3} + H_2O$$

This rearrangement is accompanied by elimination of water. The preceding reactions are carried out in acidic media, but there are other conditions that also cause rearrangement.

Now that general types of reactions found in organic chemistry have been considered, it is appropriate to investigate what lies *between* reactant and product in a typical reaction.

Intermediates

A detailed step-wise description of what happens between starting materials and product is called a mechanism. A mechanism can involve one or several intermediates. At first glance the list of possible candidates for intermediates appears hopelessly large. A closer look at the details of any organic reaction, however, should lead to one of the most exciting generalizations of the discipline. *Almost all organic reactions proceed through one of four types of intermediates.* Based on electron abundance or deficiency around the reactive carbon atom, the four types of intermediates are: a species with a positively charged carbon atom, a species with a negatively charged carbon atom, a neutral free radical with vacancy for one electron, and a neutral carbene with vacancy for two electrons.

1. A Species with a Positively Charged Carbon Atom

Carbon can carry a formal positive charge because of a deficiency of electrons. Since the species is a cation of carbon, it is called a carbocation.

$$\underset{\text{Carbocation}}{R - \underset{\underset{R}{|}}{\overset{\overset{R}{|}}{C}} \oplus}$$

6 BASIC PRINCIPLES FOR PROBLEM SOLVING IN ORGANIC CHEMISTRY

A typical S_N1 reaction, such as the hydrolysis of *t*-butyl chloride shown below, proceeds through a carbocation intermediate.

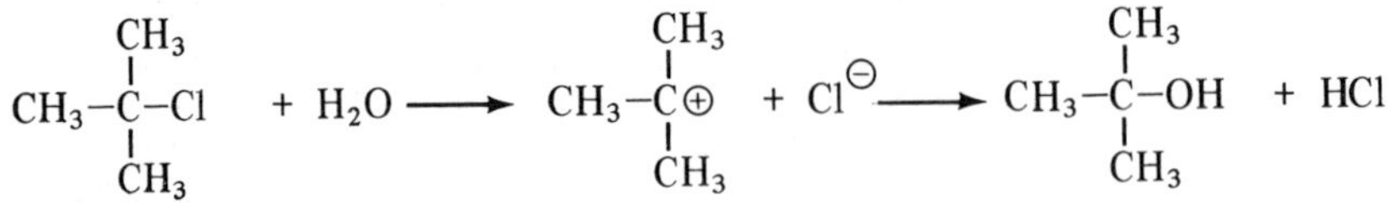

$$CH_3-\underset{CH_3}{\underset{|}{\overset{CH_3}{\overset{|}{C}}}}-Cl + H_2O \longrightarrow CH_3-\underset{CH_3}{\underset{|}{\overset{CH_3}{\overset{|}{C}}}}\oplus + Cl^\ominus \longrightarrow CH_3-\underset{CH_3}{\underset{|}{\overset{CH_3}{\overset{|}{C}}}}-OH + HCl$$

t-Butyl chloride　　　　Carbocation　　　　*t*-Butyl alcohol

However, when the charge on carbon only partially develops in an activated complex (in the transition state of a reaction), a clear-cut formal charge is not assigned.* Instead, partial charges are used to describe the complex as follows:

$$Z^{\delta\ominus}\text{-----}C^{\delta\oplus}\text{-----}X^{\delta\ominus}$$

Typically S_N2 reactions proceed through such a transition state. In the specific S_N2 example that follows iodide ion is displaced by hydroxide ion.

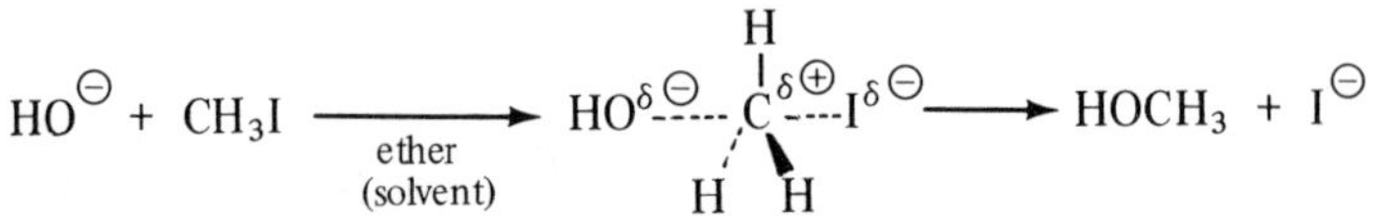

$$HO^\ominus + CH_3I \xrightarrow[\text{(solvent)}]{\text{ether}} HO^{\delta\ominus}\text{-----}\underset{H\quad H}{\overset{H}{\overset{|}{C^{\delta\oplus}}}}\text{-----}I^{\delta\ominus} \longrightarrow HOCH_3 + I^\ominus$$

No attempt is made here to equate an activated complex to an intermediate. Carbon is still positive, regardless of the degree to which the charge is developed.

2. A Species with a Negatively Charged Carbon Atom

Carbon can have a filled octet and an excess of electrons relative to the total number of protons in the nucleus. It can therefore carry a formal negative charge. This anion of carbon is appropriately named a carbanion.

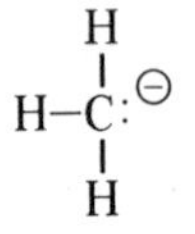

$$H-\underset{H}{\overset{H}{\overset{|}{\underset{|}{C}}}}:^\ominus$$

Carbanion

A typical reaction that proceeds through a carbanion intermediate is the Claisen condensation shown below. Once formed, the carbanion intermediate reacts with more unreacted ethyl acetate to displace ethoxide ion, $(EtO^\oplus)$, with concomitant formation of ethyl acetoacetate.

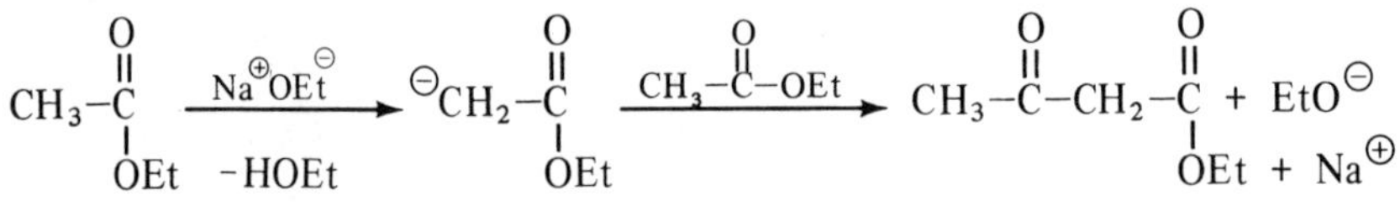

$$CH_3-\underset{OEt}{\overset{O}{\overset{\|}{C}}} \xrightarrow[-HOEt]{Na^\oplus OEt^\ominus} {}^\ominus CH_2-\underset{OEt}{\overset{O}{\overset{\|}{C}}} \xrightarrow{CH_3-\overset{O}{\overset{\|}{C}}-OEt} CH_3-\overset{O}{\overset{\|}{C}}-CH_2-\underset{OEt}{\overset{O}{\overset{\|}{C}}} + EtO^\ominus + Na^\oplus$$

Ethyl acetate　　Carbanion intermediate　　Ethyl acetoacetate

---

*Formal integer charges often do not occur in activated complexes because of unusual bonding states at high energies.

PRINCIPLES FOR DETERMINING REACTION MECHANISMS

3. **An Uncharged Carbon Atom with Vacancy for One Electron**
   In this free radical intermediate there is no charge on the carbon atom, but it is still electron deficient relative to its total electron capacity.

$$\begin{array}{c} R \\ | \\ R-C\cdot \\ | \\ R \end{array}$$

Free radical

Homolytic cleavage of a covalent bond can produce a free radical intermediate. For example, bromination of 2-methylpropane begins with homolytic dissociation of bromine and continues with the formation of a carbon-containing free radical. Product (isopropyl bromide) is formed when the isopropyl radical

Step I $Br_2 \xrightarrow{\Delta \text{ or } h\nu} 2 Br\cdot$

Step II
$$\underset{\text{2-Methylpropane}}{CH_3-\underset{\underset{CH_3}{|}}{\overset{\overset{H}{|}}{C}}-CH_3} + Br\cdot \longrightarrow \underset{\text{Isopropyl free radical}}{CH_3-\underset{\cdot}{\overset{\overset{CH_3}{|}}{C}}-CH_3} + HBr$$

Step III
$$Br_2 + \underset{\text{Isopropyl free radical}}{CH_3-\underset{\cdot}{\overset{\overset{CH_3}{|}}{C}}-CH_3} \longrightarrow \underset{\text{Isopropyl bromide}}{CH_3-\underset{\underset{Br}{|}}{\overset{\overset{CH_3}{|}}{C}}-CH_3} + Br\cdot$$

reacts with molecular bromine. Of course, there are other steps involved in the overall mechanism, but without the formation of the alkyl free radical, bromination does not occur.

4. **An Uncharged Carbon Atom with Vacancy for Two Electrons**
   Again there is no charge on the carbon atom, but it is electron deficient and divalent. Such a species is called a carbene.

Carbene

In the example that follows, dichlorocarbene is generated from chloroform and strong base and then added to cyclopentene to form the bicyclic product.

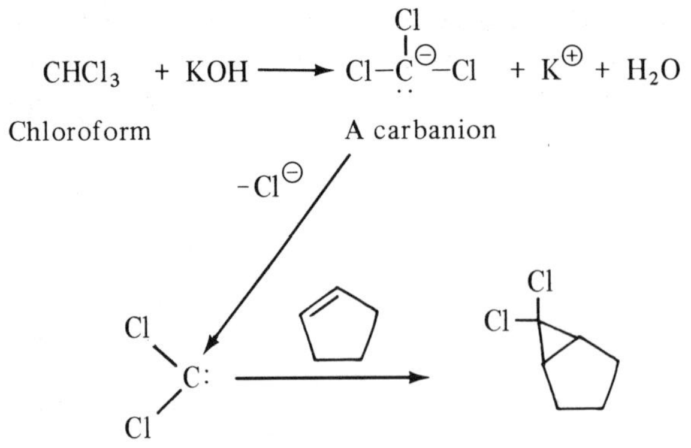

8 BASIC PRINCIPLES FOR PROBLEM SOLVING IN ORGANIC CHEMISTRY

$$CHCl_3 + KOH \longrightarrow Cl-\overset{Cl}{\underset{..}{C}}{}^{\ominus}-Cl + K^{\oplus} + H_2O$$

Chloroform   A carbanion

Dichlorocarbene → 6,6-Dichlorobicyclo[3.1.0]hexane

The choice of an intermediate is of paramount importance in writing a mechanism. Once the selection is made, each step of the proposed mechanism must be supported by fundamental principles. These principles (including polarization, inductive effects, formal charge calculation, resonance effects, guidelines for writing resonance structures, and acid–base theory) are scattered throughout leading texts. They appear here in brief form for reference and review.

Fundamental Principles

Polarization

Covalent bonds, except those between identical atoms, are polarized. The direction of polarization follows Pauling's scale of electronegativity. The most electronegative elements appear in the upper right of the Periodic Table while the most electropositive elements appear in the lower left. Accordingly, a C—Cl bond is polarized with a slight positive charge on carbon and a slight negative charge on chlorine. The opposite polarization, that is, a positive charge on chlorine and a negative charge on carbon, is not consistent with the Pauling scale. Skill at identifying the correct polarization of covalent bonds is a valuable tool for predicting sites of reactivity or determining the stability of particular organic molecules or intermediates.

Inductive Effects

Polarization can induce a dipole in a bond adjacent to the substituent or in a bond farther away. This is called the inductive effect. Inductive effects are transmitted directly through a chain of atoms within a molecule, but not across empty space or via the action of solvent molecules. These latter two descriptions apply to still another effect known as the field effect.

Inductive effects can be either electron releasing or electron withdrawing. An electron-releasing inductive effect describes the *tendency* of a section of a molecule or a substitutent to donate electrons even though formal donation of a full unit of charge

PRINCIPLES FOR DETERMINING REACTION MECHANISMS   9

does not occur. Alkyl groups are electron donors (relative to hydrogen). For example, the electron-releasing inductive effect of the methyl group appears to best account for the carbocation stability sequence of tertiary > secondary > primary > methyl.

$$CH_3 \rightarrow \underset{\underset{CH_3}{\uparrow}}{\overset{\underset{\downarrow}{CH_3}}{C^\oplus}} > CH_3 \rightarrow \underset{H}{\overset{\underset{\downarrow}{CH_3}}{C^\oplus}} > CH_3 \rightarrow \underset{H}{\overset{\underset{\downarrow}{H}}{C^\oplus}} > H - \underset{H}{\overset{\underset{\downarrow}{H}}{C^\oplus}}$$

This can be explained qualitatively by assuming there are more electrons for potential donation in the methyl group than in the hydrogen atom.

An electron-withdrawing effect refers to the *tendency* of a section of a molecule to withdraw electrons. Groups or atoms that exhibit electron-withdrawing inductive effects are more frequently encountered than those that exhibit electron-donating inductive effects. In a typical example, the bromine in bromoacetic acid withdraws electrons, making the acidic proton more easily lost than the corresponding proton in acetic acid.

$$pK_a = 2.86 \quad \text{for} \quad BrCH_2CO_2H$$

$$pK_a = 4.76 \quad \text{for} \quad CH_3CO_2H$$

The flow of electrons is toward the bromine (away from the acidic proton) in the following manner:

$$Br \leftarrow CH_2 \leftarrow \overset{\overset{O}{\|}}{C} \leftarrow O \leftarrow H$$

The anion in bromoacetate is more stable than acetate for a similar reason. The negative charge on bromoacetate is inductively stabilized by the electron-withdrawing bromine.

$$Br \leftarrow CH_2 \leftarrow \overset{\overset{O}{\|}}{C} \leftarrow \underline{\overline{O}}|^\ominus \qquad CH_3 - \overset{\overset{O}{\|}}{C} - \underline{\overline{O}}|^\ominus$$

Bromoacetate ion   Acetate ion

Calculation of Formal Charge

Formal charges resulting from the delocalization of electrons via resonance are more easily recognized than the slight centers of charge that result from inductive effects. The nature and magnitude of the formal charges can be calculated using the following formula.

$$FC = Z - (S/2 + U)$$

in which FC refers to formal charge, Z to the group number (family) in the Periodic

Table, S to the number of shared electrons, and U to the number of unshared electrons. After drawing an acceptable Lewis structure for diazomethane, CH_2N_2, for example, application of this formula results in the correct charge assignments.

$$CH_2=\overset{\oplus}{N}=\overset{\ominus}{N}$$

Redistribution of electrons in the charged diazomethane structure noted here results in additional Lewis structures for diazomethane with different locations for the formal charges.

$$\overset{\oplus}{C}H_2-\bar{N}=\overset{\ominus}{N} \qquad |\overset{\ominus}{C}H_2-\overset{\oplus}{N}\equiv N|$$

Resonance Effects

Occasionally more than one *equivalent* Lewis structure may be written for a molecule or an ion. Consider the acetate ion. There are two different Lewis structures (**I** and **II**) that can be written.

$$CH_3-C\overset{\displaystyle\bar{O}|}{\underset{\displaystyle |\underline{O}|^{\ominus}}{}} \qquad CH_3-C\overset{\displaystyle |\underline{O}|^{\ominus}}{\underset{\displaystyle \underline{O}|}{}}$$

             I   **II**

When structures differ only in the positions of electrons and not in the relative positions of the atoms, they are called resonance structures. Structures **I** and **II** are examples of resonance structures. Notice that structure **I** can be converted into structure **II** simply by changing the positions of electrons in the structure. This movement of electrons is indicated by curved arrows. *The curved arrows always indicate the direction of electron flow in the structural representations.*

$$CH_3-C\overset{\bar{O}|}{\underset{|\underline{O}|}{}} \longleftrightarrow CH_3-C\overset{|\underline{O}|^{\ominus}}{\underset{\underline{O}|}{}}$$

The structures themselves are connected by double-headed arrows (⟷). These double-headed arrows serve as reminders of the hypothetical nature of the resonance structures. That is, resonance structures do not equilibrate with each other. Instead, they represent our attempts as chemists to describe an entity that really is a hybrid of more than one Lewis structure.

A better description of the acetate ion using a single representative structure is the following:

$$CH_3-C\overset{O^{\delta^- 1/2}}{\underset{O^{\delta^- 1/2}}{}}$$

PRINCIPLES FOR DETERMINING REACTION MECHANISMS 11

The charges on each oxygen atom are identical; both carbon–oxygen bond lengths are equal. In practice we could write the non-Lewis structure here in an attempt to represent the hybrid, or we could write all the formal Lewis resonance structures for the acetate ion and then mentally remind ourselves of the hybrid. The latter approach is used in the examples that follow.

Two equivalent Lewis structures for benzene can be produced via the redistribution of electrons.

The mental *reminder* from the double-headed arrows is of a hybrid in which all the carbon–carbon bonds are of equal length, a structure symbolically represented by:

Similarly, the nitrate ion can be represented by three equivalent Lewis structures.

These structures symbolize a hybrid in which the negative charge is distributed equally over all the oxygen atoms as shown in the following hybrid structure.

Rules for Writing Resonance Structures

The practice of writing different Lewis structures by changing the positions of electrons (a process known as writing resonance structures) is guided by a set of rules:

1. Only bona fide Lewis structures are permitted. These structures result from the *proper valence electron contribution from each element involved in the molecule*. So, a neutral carbon atom may not have five bonds, a neutral hydrogen atom may not have two, and so on.
2. The position of the atoms must not change; only the *positions* of electrons may vary. Notice that the change in the position of only one hydrogen atom results in an entirely different compound, an isomer. This process is known as isomerization, not resonance.

12 BASIC PRINCIPLES FOR PROBLEM SOLVING IN ORGANIC CHEMISTRY

$$CH_3-\underset{\underset{O}{\parallel}}{C}-CH_3 \rightleftarrows CH_2=\underset{\underset{OH}{|}}{C}-CH_3 \text{ (an isomer)} \quad (1)$$

A change in the position of electrons, however, provides an acceptable resonance structure.

$$CH_3-\underset{\underset{O}{\parallel}}{C}-CH_3 \longleftrightarrow CH_3-\underset{\underset{O^{\ominus}}{|}}{\overset{\oplus}{C}}-CH_3 \quad (2)$$

Move only electrons when changing from one resonance structure to another.

3. Coplanarity of the atoms involved is a requirement for resonance. This coplanarity is necessary in order to maximize the p orbital overlap of the atoms in the structures represented by the resonance hybrid.

4. The number of unpaired electrons must not change from structure to structure. For example, $\dot{C}H_2N=\dot{N}|$ would not be allowed for diazomethane, CH_2N_2, since it shows two unpaired electrons.

5. The energy of the actual molecule is lower than for any individual resonance structure. Specifically, the hybrid for the acetate ion is more stable (lower in energy) than either of the two Lewis resonance structures.

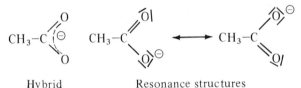

Hybrid   Resonance structures

6. All resonance structures do not contribute equally to the overall molecular description. For example, even though it is possible to write the following resonance structure for acrolein ($CH_2=CH-CH=O$).

$$|\overset{\ominus}{C}H_2-CH=CH-\overset{\oplus}{O}|$$

it is unreasonable to think that it makes a substantial contribution to the complete description since the charges are opposite those predicted from the electronegativities of carbon and oxygen. In addition, structures leading to a large charge separation, such as the one just shown, are disfavored.

Acid–Base Theory

Brønsted Theory. An acid in the Brønsted sense is anything that can donate a proton; a base is anything that can accept a proton. The underlined proton in $CH_3CO_2\underline{H}$ is acidic; it is part of a functional group, $-CO_2H$, known as a carboxylic acid. Dissociation of a carboxylic acid occurs in water to produce the conjugate base of the acid and the hydronium ion. For example, acetic acid ionizes in water to produce acetate ion and hydronium ion.

$$CH_3CO_2H + H_2O \rightleftharpoons CH_3CO_2^{\ominus} + H_3O^{\oplus}$$

Acetic acid　　　　　　Acetate ion　Hydronium ion

Identification of the acidic hydrogen in this example is an easy task because of the name of the functional group, but recognition of the acidic protons on molecules (or parts of the same molecule), which do not contain the carboxylic acid group requires more effort.

In the absence of a carboxylic acid group, a loosely held proton often can be identified by considering the stability of the conjugate base of the potential acid. If two or more resonance structures of nearly equivalent energy can be written for the conjugate base, then a reasonable case can be made for characterizing the proton on the original acid as acidic.

For example, the conjugate base of phenol, phenoxide ion, is stabilized by the phenyl ring in the following way:

Phenoxide ion

Similarly, the conjugate base of acetone is stabilized in a fashion that could not occur for the conjugate base of a hydrocarbon such as methane (methide ion, CH_3^{-1}).

Acetone　　　　　　　　　Resonance stabilization

Methane　　　　Methide ion
　　　　　　　(no stabilization)

In another example, 2,4-pentanedione, a very acidic proton resides on the number three carbon atom because the two carbonyl groups stabilize the conjugate base.

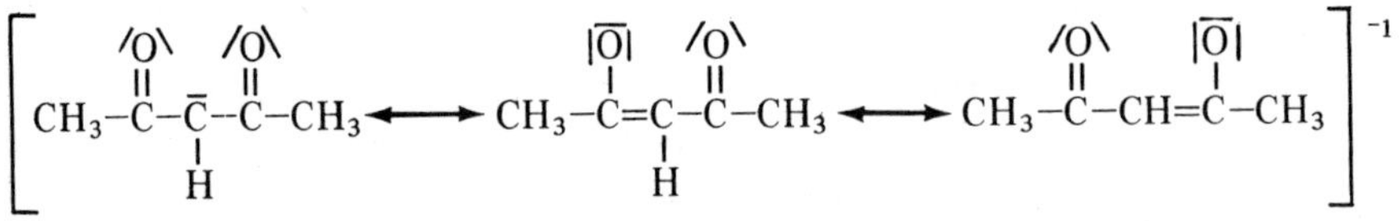

$$\left[CH_3-\overset{\overset{\displaystyle O}{\|}}{C}-\overset{\overset{\displaystyle}{|}}{\underset{\underset{\displaystyle H}{|}}{C}}-\overset{\overset{\displaystyle O}{\|}}{C}-CH_3 \longleftrightarrow CH_3-\overset{\overset{\displaystyle \bar{O}}{|}}{C}=\underset{\underset{\displaystyle H}{|}}{C}-\overset{\overset{\displaystyle O}{\|}}{C}-CH_3 \longleftrightarrow CH_3-\overset{\overset{\displaystyle O}{\|}}{C}-CH=\overset{\overset{\displaystyle \bar{O}}{|}}{C}-CH_3 \right]^{-1}$$

In all three of these examples, the existence of a viable resonance-stabilized conjugate base supports the choice of the acidic proton.

The evaluation of the acidity of an organic compound involves consideration of both resonance and inductive effects. Groups bonded directly to the atom bearing the proton exert the strongest inductive effects. The proton is tightly or loosely held depending in part on these inductive effects. Once the proton is lost, however, resonance effects become important in contributing to the stability of the newly formed conjugate base. Thus, both effects significantly influence acidity.

Nitromethane (CH_3NO_2) for example, is acidic due to the inductive effect of the nitro group. Once formed, the anion is stabilized by resonance. Part of this latter stabilization is available in the transition state for proton loss, thus enhancing the rate of ionization.

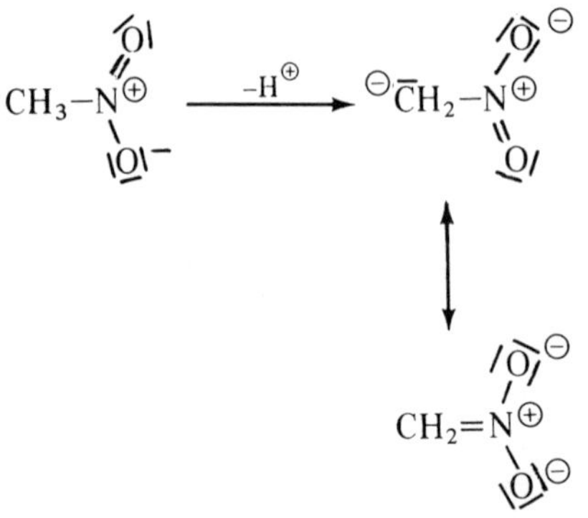

In summary, once the acidic proton has been identified, acidity will depend on the ease of proton abstraction and the amount of stabilization in the conjugate base.

Lewis Theory. All the examples of acids mentioned in the preceding discussions have a loosely held proton. In the absence of such a proton, it is still possible to describe a compound as an acid by employing the Lewis theory. The Lewis theory complements the Brønsted theory by defining an acid as an acceptor of an electron pair and base as the donor of an electron pair.

Now that the definition of an acid has been extended, which approach is better used to characterize a given reaction? First, as a practical consideration, it is useful to try to characterize the reaction according to the Brønsted theory. Then, if no loosely held proton is available, the Lewis theory should be employed. For example, attempts to describe the following as a Brønsted acid–base reaction fail.

PRINCIPLES FOR DETERMINING REACTION MECHANISMS

$$Cl_3Al + :Cl^{\ominus} \longrightarrow Cl_3Al:Cl^{\ominus}$$

However, according to Lewis theory, aluminum chloride is the acid and the chloride ion is the base. Similarly, in a typical addition reaction, the alkene with its electron rich double bond is the Lewis base and the electrophile, BH_3, is the Lewis acid.

$$R_2C=CR_2 + BH_3 \longrightarrow \underset{\underset{H}{\overset{|}{H-B}}}{\overset{R}{\underset{|}{\overset{|}{R-C}}}}-\underset{H}{\overset{R}{\underset{|}{\overset{|}{C-R}}}}$$

In a final example of the application of Lewis theory, the acid and base are BF_3 and NH_3, respectively.

$$BF_3 \;+\; :NH_3 \longrightarrow \overset{\oplus\;\;\;\ominus}{H_3N-BF_3}$$

Lewis acid   Lewis base   Salt

Regardless of which theory is invoked, basicity is a function of the availability of the electron pair. This availability depends upon the inductive and/or resonance effects of groups bonded directly to the atom bearing the free electron pair. Electron-withdrawing groups decrease the availability of the electron pair and hence the basicity.

$$O_2N{-}C_6H_4{-}\ddot{N}H_2 \;<\; C_6H_5{-}\ddot{N}H_2 \;<\; :NH_3$$

⎯⎯⎯⎯⎯⎯⎯⎯⎯⎯⎯⎯⎯⎯⎯⎯⎯⎯⎯⎯⎯⎯→
basicity increases

Electron-donating groups increase the basicity.

$$\ddot{N}H_3 \;<\; CH_3{\rightarrow}\ddot{N}H_2 \;<\; CH_3{\rightarrow}\ddot{N}H{\leftarrow}CH_3 \;<\; CH_3{\rightarrow}\underset{\uparrow CH_3}{\overset{\downarrow CH_3}{N:}}{\leftarrow}CH_3$$

⎯⎯⎯⎯⎯⎯⎯⎯⎯⎯→
basicity increases

In short, these acid–base theories reinforce each other; they are not mutually exclusive. A base always contains a free electron pair in either theory, but the list of substances classified as acids expands considerably with the Lewis theory. The emphasis shifts from the proton in the Brønsted theory, to the electron pair in the Lewis theory.

16 BASIC PRINCIPLES FOR PROBLEM SOLVING IN ORGANIC CHEMISTRY

A Few Guidelines for Writing Organic Reaction Mechanisms

1. A mechanism must explain the formation of product.
2. Typically, isolable organic compounds (not intermediates) contain four bonds. Carbon has four bonds (only four) in its ground state.
3. Curved arrows indicate the *direction* of electron flow within a structure.
4. The strongest base that can exist in water is the hydroxide ion; the strongest acid that can exist in water is the hydronium ion, $H_3O^{\oplus}$.
5. In an acidic solution hydroxide ion is never lost from an alcohol. Instead, hydroxide ion departs as a water molecule. For example,

$$R-OH \xrightleftharpoons{H^{\oplus}} R-\overset{H}{\underset{\oplus}{O}}H \longrightarrow R^{\oplus} + HOH$$

6. The order of carbocation stability is

    tertiary > secondary > primary

7. The order of free radical stability is

    tertiary > secondary > primary

8. The order of carbanion stability is

    primary > secondary > tertiary

9. Carbonyl carbon atoms, C=O, are electropositive; carbonyl oxygen atoms are electronegative.
10. In general, resonance effects are more important than inductive effects in stabilizing intermediates.

2 PROBLEMS

*1. What formal charges, if any, are found on each atom in the following?

(a) HNO_3

(b) 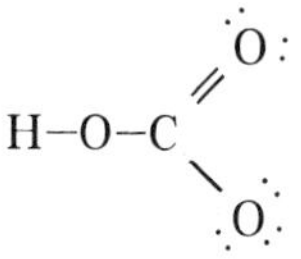 H—O—C(=O:)(—O:)

(c) CH_3—N(=O:)(—O:)

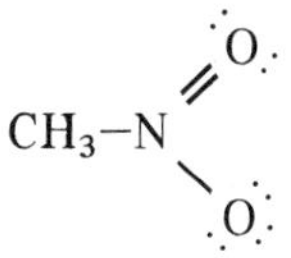

(d) 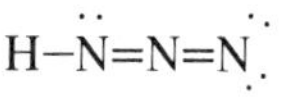 H—N=N=N:

Chapter 1
General Structure

*2. Determine the empirical formula of an organic compound whose percentage composition determined by combustion analysis is:

40.01% carbon   6.11% hydrogen

If the molecular mass of the compound is 180, what is the molecular formula?

Chapters 1 and 2
Formula Determination

*3. State whether the structures of each pair represent the same compound or different compounds.

(a) CH_3CH_2Cl and   H—C(H)(H)—C(H)(H)—Cl

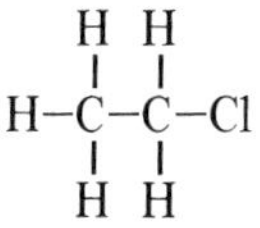

(b) 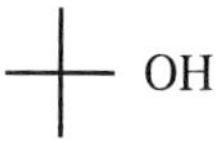 ⊢—OH   and $(CH_3)_3C$—OH

(c) 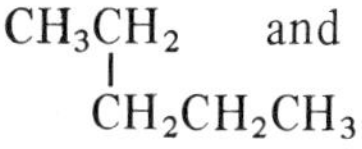 CH_3CH_2—$CH_2CH_2CH_3$ and $CH_3CH_2CH_2CH_2CH_3$

(d) 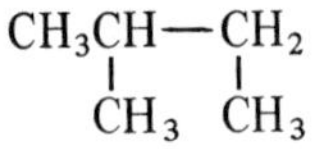 CH₃CH—CH₂ and structure of 2,3-dimethylbutane drawn out
      | |
      CH₃ CH₃

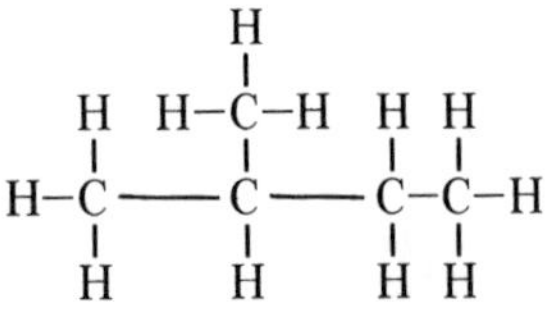

(e) CH₃CH₂OCH₃ and H-C(H)(H)-C(=O)-C(H)(H)-H (propanal drawn out)

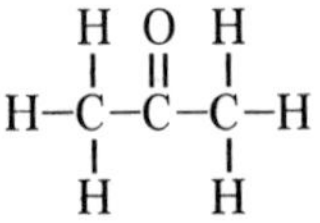

(f) CH₃OCH₂CH₃ and CH₃CH₂OCH₃

Chapters 1 and 2
Isomerism

*4. Discuss the geometry about carbon atoms in the following compounds:

(a) CH₂=CHCH₃

(b) CH₃C≡CH

(c) CH₃CH₂CH₃

Chapters 1 and 2
Bonding in Hydrocarbons

*5. Show the net dipole moment in the following molecules:

BeCl₂, BF₃, NH₃, H₂O, and (CH₃)₂O

Chapter 1
Shapes of Molecules

*6. How are the structures related within each pair?

(a) 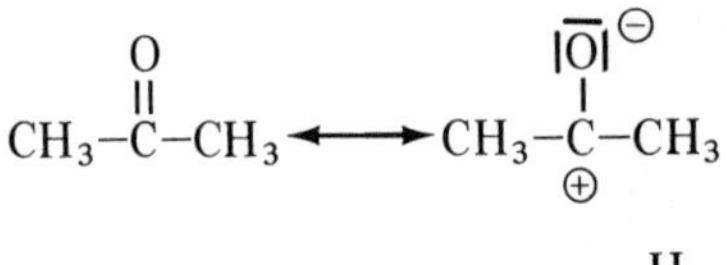 CH₃−C(=O)−CH₃ ⟷ CH₃−C(−Ö:⁻)(⊕)−CH₃

(b) 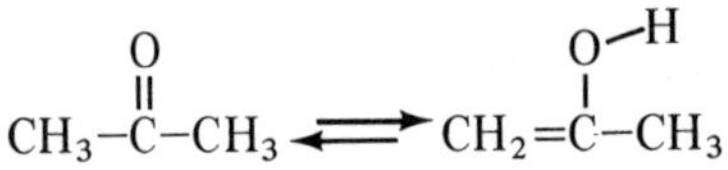 CH₃−C(=O)−CH₃ ⇌ CH₂=C(−O−H)−CH₃

Chapters 1 and 2
Isomerism versus Resonance

*7. In the following pairs, which compound should have the higher boiling point? Why?

(a) CH$_3$(CH$_2$)$_2$–C(=O)–OH   or   CH$_3$CH$_2$–C(=O)–OCH$_3$
    Butanoic acid   or   Methyl propanoate

(b) CH$_3$CH$_2$–O–CH$_2$CH$_3$ or   CH$_3$CH$_2$CH$_2$CH$_2$–OH
    Diethylether   or   1-Butanol

(c) (CH$_3$CH$_2$)$_3$N   or   CH$_3$(CH$_2$)$_5$NH$_2$
    Triethylamine   or   Hexylamine

(d) CH$_3$CH$_2$CH$_2$CH$_3$ or   CH$_3$(CH$_2$)$_6$CH$_3$
    Butane   or   Octane

(e) CH$_3$CH$_2$C(=O)CH$_3$ or   CH$_3$CH$_2$CH$_2$CH$_2$–OH
    Butanone   or   1-Butanol

(f) cis-2-Butene   or   trans-2-Butene

Chapter 2
Representative Carbon Compounds

*8. Diethyl ether (CH$_3$CH$_2$–O–CH$_2$CH$_3$) dissolves in concentrated sulfuric acid (H$_2$SO$_4$) but not in water. Explain.

Chapter 2
Representative Carbon Compounds and Ethers

*9. Which has the higher pK_a? Why?

CH$_3$CO$_2$H   or   CF$_3$CO$_2$H

Chapter 2
General Acidity

*10. Draw all the possible isomers for C$_3$H$_8$O.

Chapter 2
Isomerism

20 PROBLEMS

*11. How many different ethers can you draw for $C_6H_{14}O$?

Chapter 2
Isomerism

*12. Draw an acceptable Lewis structure for the acetate ion, $CH_3CO_2^{\ominus}$. Account for the fact that both carbon to oxygen bonds are equal in length.

Chapter 2
Resonance

*13. Draw the structures for all the alkyl chlorides with the formula C_4H_9Cl.

Chapter 3
Isomerism

*14. Show the arrangement of the chlorine atom (axial or equatorial) in the following compound. Explain why this conformation is preferred.

Chapter 3
Cycloalkanes

*15. Starting with ethane, cyclohexane, and any inorganic reagents, synthesize ethylcyclohexane.

Chapter 3
Synthesis of Alkanes

*16. Name the following compounds:

(a)   (b)

(c)   (d)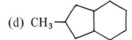

Chapter 3
Bicyclic Compounds

PROBLEMS 21

*17. Name the following alcohols.

(a) CH₃CH₂CH₂CH₂OH

(b) CH₃C(CH₃)(CH₂CH₃)CH₂CH₂OH

(c) CH₃CH(OH)CH₂CHCH₃ ... CH₃CH(CH₃)CH₂CH(OH)CH₃

(d) CH₃CHCH₂C(CH₃)₂—CH(CH₃)CH₃ with OH

**Chapter 3
Nomenclature of Alcohols**

*18. Name the following hydrocarbons:

(a) CH₃CH(CH₃)CH₂C(CH₃)₂CH₃

(b) CH₃CH₂CH₂CH(CH₃)CH(CH₃)... with two CH₃ branches

(c) CH₃CH(CH₂CH₃)—CH₂CH₂CH(CH₂CH₂CH₃)CH₂CH₃

(d) CH₃CH₂—C(CH₃)(CH₂CH₃)—CH₂CH₂CH₃

**Chapter 3
Nomenclature**

*19. Draw a three-dimensional conformational representation of *trans*-decalin, bicyclo[4.4.0] decane.

**Chapter 3
Conformational Analysis**

**20. How many different dichlorocyclopentanes can result from the chlorination of cyclopentane? Consider both stereoisomers and geometric isomers.

**Chapters 3 and 4
Cycloalkanes**

22 PROBLEMS

*21. A compound of formula C_5H_{10} does not decolorize potassium permanganate. Monochlorination of this compound can yield only *one possible product*. What is the structure of the compound?

Chapters 3 and 4
Alkanes

**22. What is the detailed mechanism for the chlorination of ethane?

$$CH_3CH_3 + Cl_2 \xrightarrow{h\nu} CH_3CH_2Cl + HCl$$

Chapter 4
Alkanes

**23. Consider the following two reaction coordinate diagrams. One represents reaction of a bromine atom with methane; the other represents homolytic dissociation of Br_2. Label the axes of these diagrams. Indicate what species are at the Points A, B, C, D, and E and what distances represent ΔH^0 and E_{act} in each case.

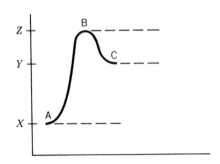 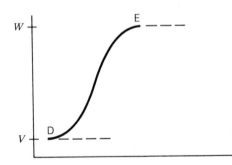

Chapter 4
Alkanes

*24. Why is the formation of isopropyl bromide favored over propyl bromide in the following reaction?

$$CH_3CH_2CH_3 \xrightarrow[\substack{\Delta \\ (127°)}]{Br_2} \underset{\underset{97\%}{Br}}{CH_3\overset{|}{C}HCH_3} + \underset{3\%}{CH_3CH_2CH_2Br} + HBr$$

Chapter 4
Bromination of Alkanes

*25. Draw structures for all the monochlorination products of 2-methylbutane.

$$CH_3CHCH_2CH_3 \xrightarrow[300°C]{Cl_2} ?$$
$$|$$
$$CH_3$$

Would you expect different products from bromination? Why?

$$CH_3CHCH_2CH_3 \xrightarrow[127°C,\ h\nu]{Br_2} ?$$
$$|$$
$$CH_3$$

Chapter 4
Reactions of Alkanes

*26. What is the major product of the following reaction? Why is this the only product?

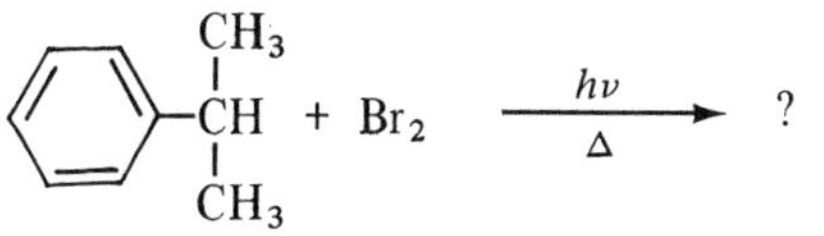

$$C_6H_5-CH(CH_3)_2 + Br_2 \xrightarrow[\Delta]{h\nu} ?$$

Chapter 4
Alkanes

*27. Arrange in order of decreasing nucleophilicity.

$$CH_3CO_2^{\ominus} \quad H_2O \quad CH_3OH \quad CH_3O^{\ominus} \quad HO^{\ominus}$$

Chapter 5
Substitution Reactions

**28. What are the products of the following reaction? What is the reaction mechanism?

$$C_6H_5-\underset{\underset{CH_3}{|}}{\overset{\overset{C_6H_5}{|}}{C}}-H + FSO_3H \cdot SbF_5 \longrightarrow ?$$

Chapter 5
Substitution Reactions

*29. Consider the following data for a simple displacement reaction:

$$^{\ominus}OH + R-Br \longrightarrow R-OH + Br^{\ominus}$$

Experiment Number	Initial Concentration (moles/L)		Relative Initial Reaction Rate
	$^{\ominus}OH$	RBr	
1	0.05	0.05	1
2	0.10	0.05	1
3	0.05	0.10	2

What is the mechanism of this reaction?

Chapter 5
Substitution Reactions and
Elimination Reactions

PROBLEMS

*30. Show the product and mechanism for the following solvolysis reaction?

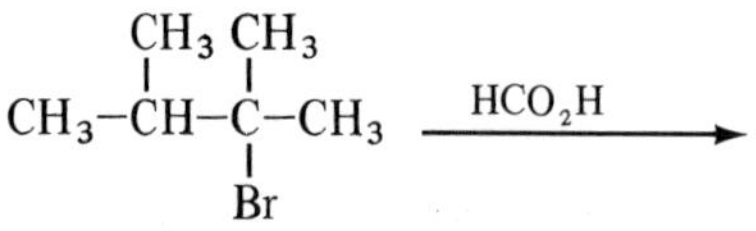

$$\text{CH}_3-\underset{\text{H}}{\overset{\text{CH}_3}{\text{CH}}}-\underset{\text{Br}}{\overset{\text{CH}_3}{\text{C}}}-\text{CH}_3 \xrightarrow{\text{HCO}_2\text{H}}$$

Chapter 5
Substitution Reactions

*31. What are the product and mechanism for the following?

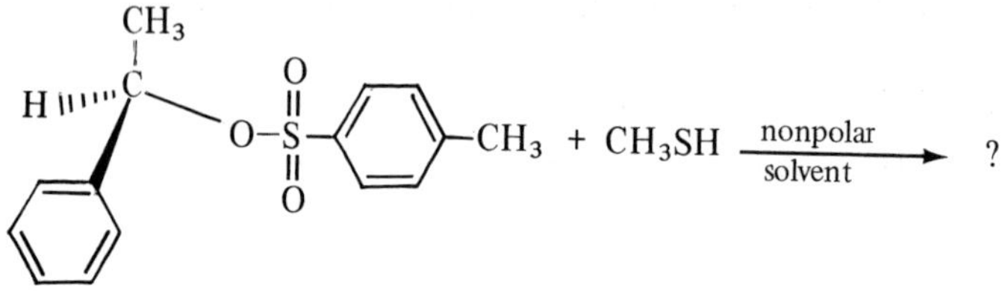

(structure: chiral carbon bearing H, CH₃, phenyl, and O–S(=O)₂–C₆H₄–CH₃) + CH₃SH $\xrightarrow[\text{solvent}]{\text{nonpolar}}$?

Chapter 5
Substitution Reactions

*32. What is the product of the following reaction? What is the mechanism for this reaction?

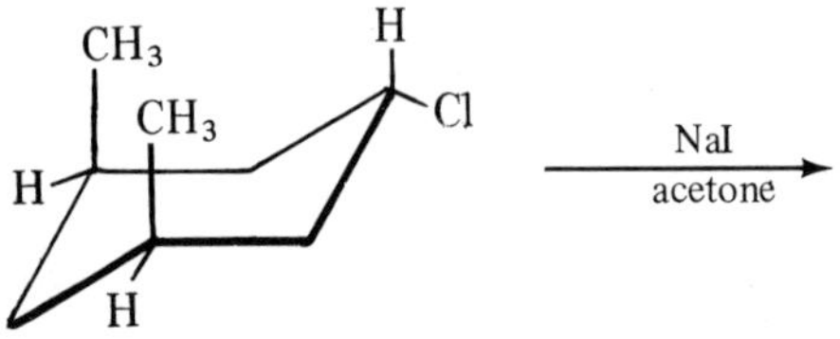

(cyclohexane with CH₃, CH₃ (gem-dimethyl) and Cl substituents) $\xrightarrow[\text{acetone}]{\text{NaI}}$

Chapters 3 and 5
Substitution Reactions

*33. What are the products of the following two reactions? What are the mechanisms and what common function is carried out by Ag⁺ in Reaction (1) and H⁺ in reaction (2)?

(1) $\quad \text{CH}_3\text{CH}_2\text{Br} + \text{CH}_3\text{CH}_2\text{OH} \xrightarrow[\Delta]{\text{Ag}^{\oplus}}$

(2) $\quad \text{CH}_3\text{CH}_2\text{OH} + \text{HBr} \xrightarrow{\Delta}$

Chapter 5
Substitution Reactions

*34.

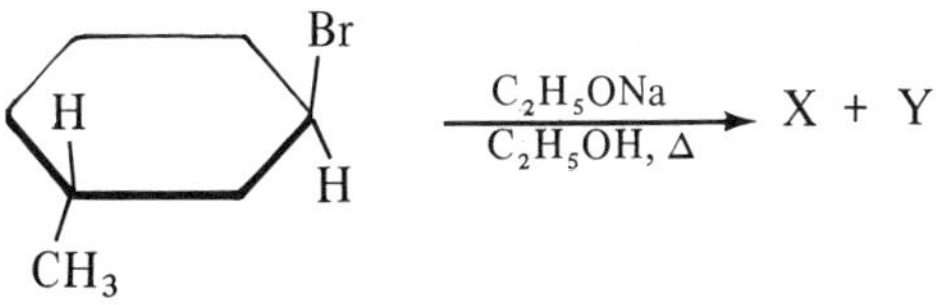

[Structure: cyclohexane ring with Br and H on one carbon (trans), H on adjacent carbon, and CH₃ on another position]

$$\xrightarrow[C_2H_5OH, \Delta]{C_2H_5ONa}$$ X + Y (both have the same empirical formula)

What is the mechanism of this reaction? What are X and Y and what is the relationship between them? (Presume that nucleophilic substitution does not occur.)

Chapter 5
Substitution Reactions and
Elimination Reactions

*35. What are the structures of three products that can result from putting *t*-butyl chloride in a mixture that is 80% methanol and 20% water? What are the mechanistic routes to these products?

Chapter 5
Substitution Reactions

*36. Show the mechanism of formation for the *major* product.

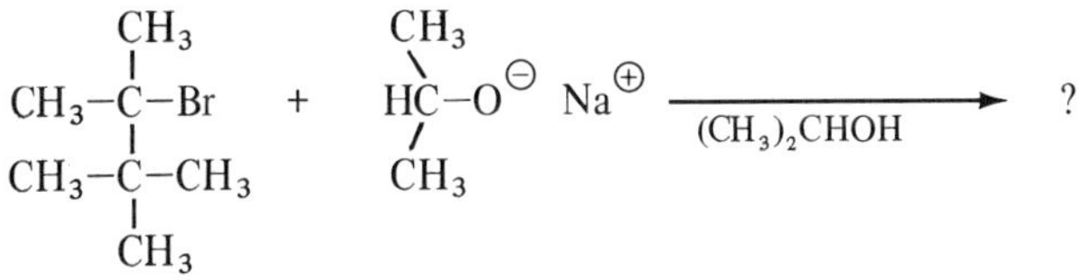

$$\underset{\underset{\underset{CH_3}{|}}{\overset{\overset{CH_3}{|}}{CH_3-C-CH_3}}}{\overset{\overset{CH_3}{|}}{CH_3-C-Br}} \quad + \quad \underset{CH_3}{\overset{CH_3}{\diagdown}}HC-O^{\ominus}\ Na^{\oplus} \quad \xrightarrow{(CH_3)_2CHOH} \quad ?$$

Chapter 5
Substitution Reactions

*37. What is the product of the following reaction?

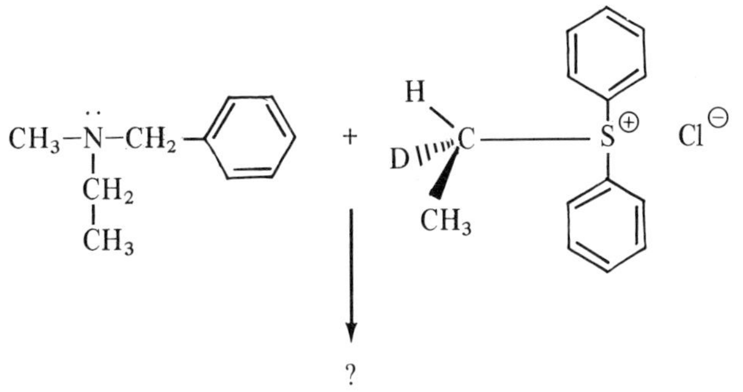

$$CH_3-\underset{\underset{CH_3}{\overset{|}{CH_2}}}{\overset{..}{N}}-CH_2-C_6H_5 \quad + \quad \underset{CH_3}{\overset{H}{D^{\,\prime\prime\prime}C}}-\overset{\oplus}{S}(C_6H_5)_2\ Cl^{\ominus}$$

$$\downarrow$$

?

Chapter 5
Substitution Reactions

26 PROBLEMS

*38. Why is the product ratio (of alcohol to ether) expected to be about the same in each of the following reactions, whereas the rate of each reaction differs?

(1) 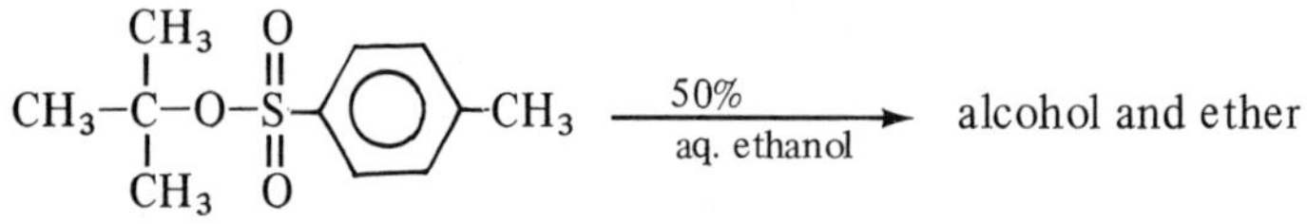 $CH_3-\underset{\underset{CH_3}{|}}{\overset{\overset{CH_3}{|}}{C}}-O-\underset{\underset{O}{\|}}{\overset{\overset{O}{\|}}{S}}-C_6H_4-CH_3 \xrightarrow[\text{aq. ethanol}]{50\%}$ alcohol and ether

(2) 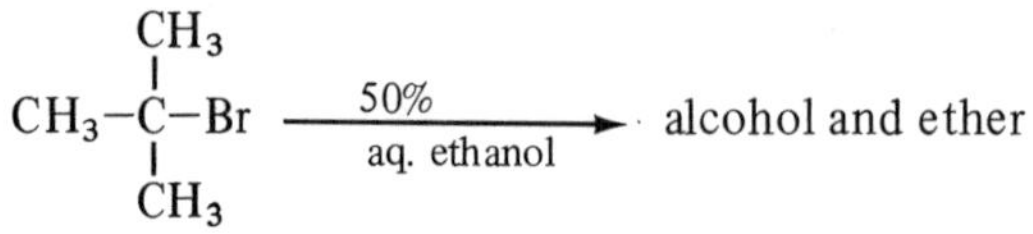 $CH_3-\underset{\underset{CH_3}{|}}{\overset{\overset{CH_3}{|}}{C}}-Br \xrightarrow[\text{aq. ethanol}]{50\%}$ alcohol and ether

Chapter 5
Substitution Reactions

**39. What is a reasonable structure for the product? How does the reaction occur?

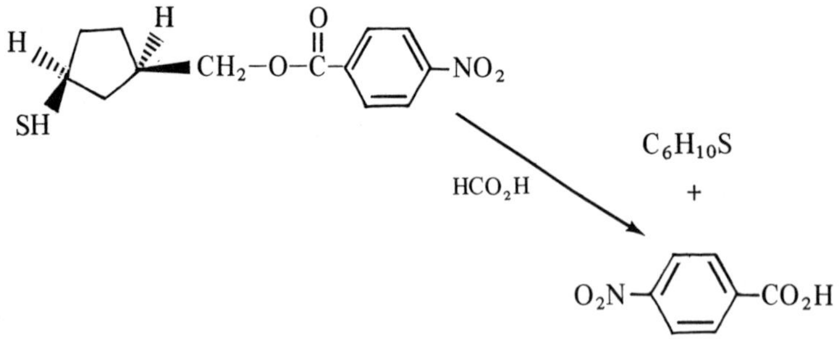

$$\text{(cyclopentane with } H, H, SH, \text{ and } CH_2-O-\overset{O}{\overset{\|}{C}}-C_6H_4-NO_2 \text{ substituents)} \xrightarrow{HCO_2H} C_6H_{10}S \;+\; O_2N-C_6H_4-CO_2H$$

Chapter 5 and Special
Topic H
Substitution Reactions

*40. Which of the following is the most stable cabocation and why?

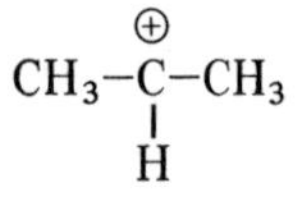

A: $CH_3-\overset{\oplus}{\underset{\underset{H}{|}}{C}}-CH_3$

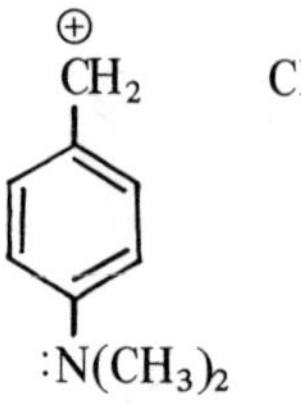

B: $\overset{\oplus}{C}H_2-C_6H_4-:N(CH_3)_2$

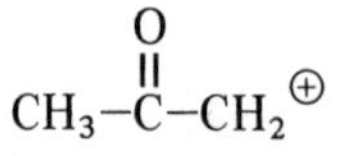

C: $CH_3-\overset{\overset{O}{\|}}{C}-\overset{\oplus}{C}H_2$

Chapters 5 and 6
Carbocations

41.

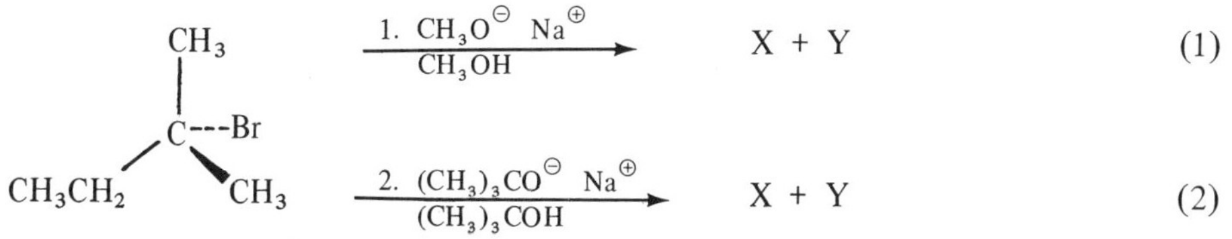

$$\underset{CH_3CH_2}{\overset{CH_3}{\underset{|}{C}}}\overset{Br}{\underset{CH_3}{}}$$

$\xrightarrow[CH_3OH]{1.\ CH_3O^{\ominus}\ Na^{\oplus}}$ X + Y (1)

$\xrightarrow[(CH_3)_3COH]{2.\ (CH_3)_3CO^{\ominus}\ Na^{\oplus}}$ X + Y (2)

What are the products X and Y for these reactions? Which is the major product in each case? What is the reason for your answer? Neither product is an ether.

Chapters 5 and 6
Substitution Reactions and
Elimination Reactions

42.

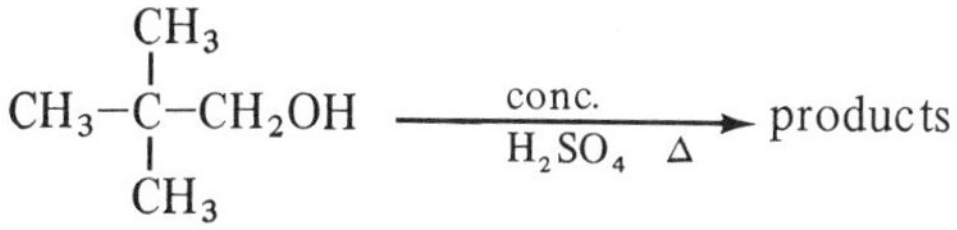

$$CH_3-\underset{\underset{CH_3}{|}}{\overset{\overset{CH_3}{|}}{C}}-CH_2OH \xrightarrow[H_2SO_4\ \Delta]{conc.} products$$

Show the products and write a mechanism for the reaction.

Chapters 5 and 6
Alkenes

*43. Propose a structure for a hydrocarbon, C_7H_{12}, which is converted to C_7H_{14} by catalytic hydrogenation. How many rings must the structure contain?

Chapter 6
Alkenes and Cycloalkenes

*44. Describe in detail the motions you must use to convert the structure on the left to the (same) structure on the right.

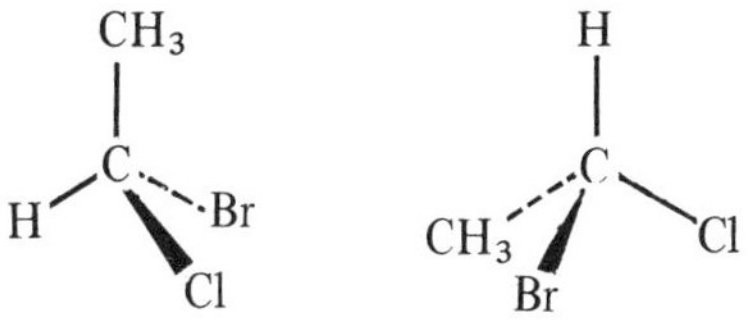

Chapters 3 and 6
Stereochemistry

**45. Show the reagents needed to accomplish the following sequence. Explain what is occurring at each step.

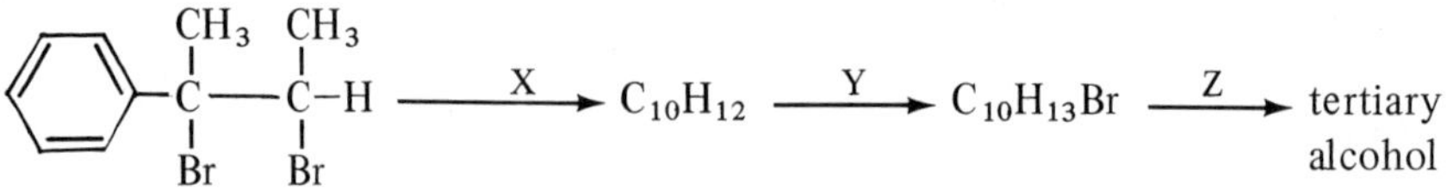

28 PROBLEMS

$$\text{Ph}-\underset{\underset{\text{Br}}{|}}{\overset{\overset{\text{CH}_3}{|}}{\text{C}}}-\underset{\underset{\text{Br}}{|}}{\overset{\overset{\text{CH}_3}{|}}{\text{C}}}-\text{H} \xrightarrow{X} C_{10}H_{12} \xrightarrow{Y} C_{10}H_{13}Br \xrightarrow{Z} \text{tertiary alcohol}$$

Chapters 5–7
Alkenes

*46. Draw structural formulas for the products of the following reactions:

(a) 1-pentene + HCl $\xrightarrow{\Delta}$

(b) 1-pentene + (conc.)H$_2$SO$_4$ $\xrightarrow{\text{cold}}$

(c) 1-pentene + HBr $\xrightarrow{\text{peroxides}}$

(d) 1-pentene + O$_3$ $\xrightarrow{\text{Zn,H}_2\text{O}}$

(e) 1-pentene + Br$_2$ + H$_2$O $\longrightarrow$

Chapter 7
Alkenes

***47. Show a mechanism for the following reaction, and indicate where the deuteriums may be in the products:

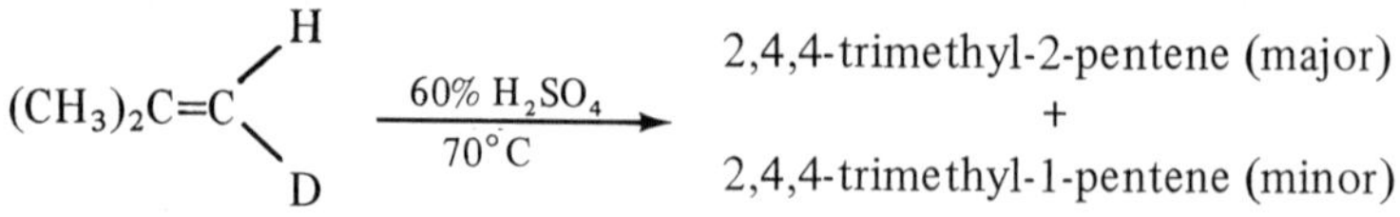

$$(CH_3)_2C=C\underset{D}{\overset{H}{\diagdown\!\!\!\diagup}} \xrightarrow[70°C]{60\% \text{ H}_2\text{SO}_4} \begin{array}{l} \text{2,4,4-trimethyl-2-pentene (major)} \\ + \\ \text{2,4,4-trimethyl-1-pentene (minor)} \end{array}$$

Chapter 7
Alkenes

**48. What are the products X and Y for the following reaction sequence and what is a reasonable mechanism for the indicated processes?

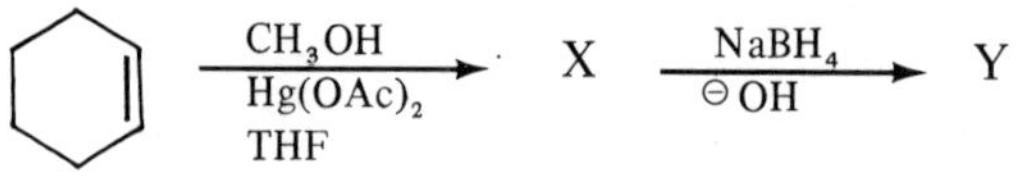

$$\bigcirc \xrightarrow[\text{THF}]{\underset{\text{Hg(OAc)}_2}{\text{CH}_3\text{OH}}} X \xrightarrow[\ominus \text{OH}]{\text{NaBH}_4} Y$$

Chapter 7
Alkenes

*49.
$$\underset{CH_3}{\overset{CH_3}{\diagdown\!\!\!\diagup}}C=CH_2 \xrightarrow{(BH_3)_2} X \xrightarrow[\ominus \text{OH}]{H_2O_2} Y$$

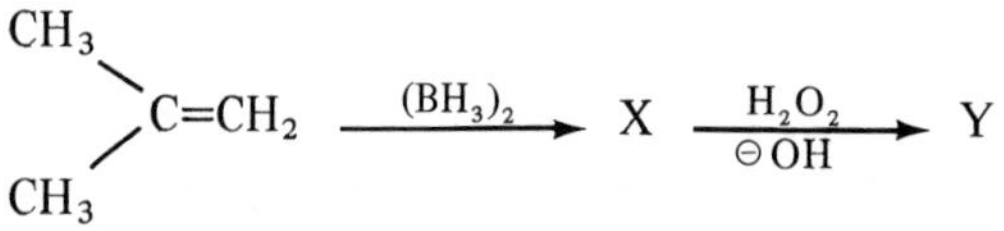

What are X and Y and what is the mechanism for the formation of X?

Chapter 7
Alkenes

*50. What are the reagents A, B, and C?

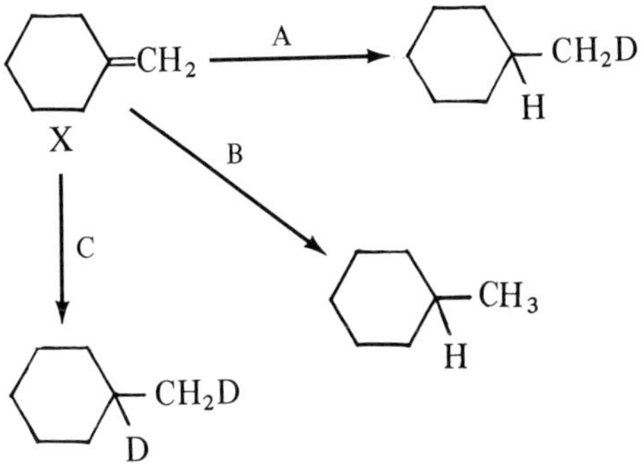

**Chapter 7
Alkenes**

*51. Draw the structures for the alkenes that form the following ozonization products when treated with ozone and water.

(a) cyclopentane-1,3-dione, CH₃CHO, CH₃-CO-CH₃

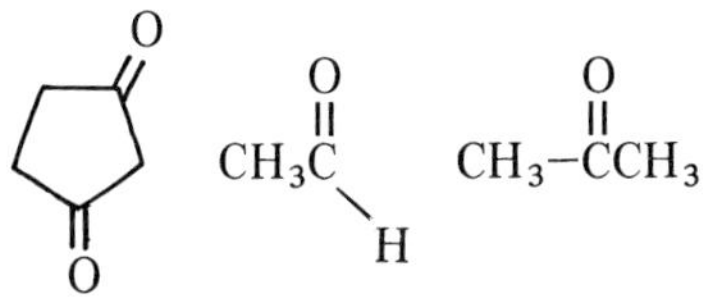

(b) decalin-1,6-dione (bicyclic diketone)

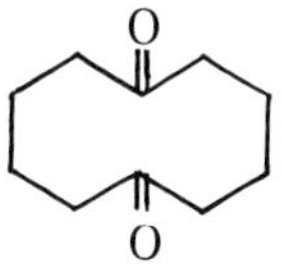

(c) cyclopentanone

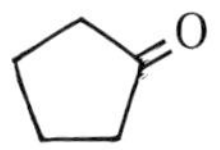

**Chapter 7
Alkenes**

***52. Show the mechanism and the products of the following reaction. Indicate clearly the stereochemical relationships among the products.

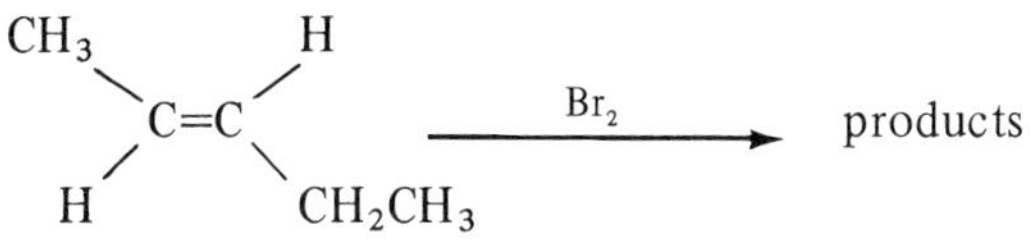

$$\text{(CH}_3\text{)(H)C=C(H)(CH}_2\text{CH}_3\text{)} \xrightarrow{\text{Br}_2} \text{products}$$

**Chapters 7 and 8
Alkenes**

30 PROBLEMS

*53. Write structural formulas for the products.

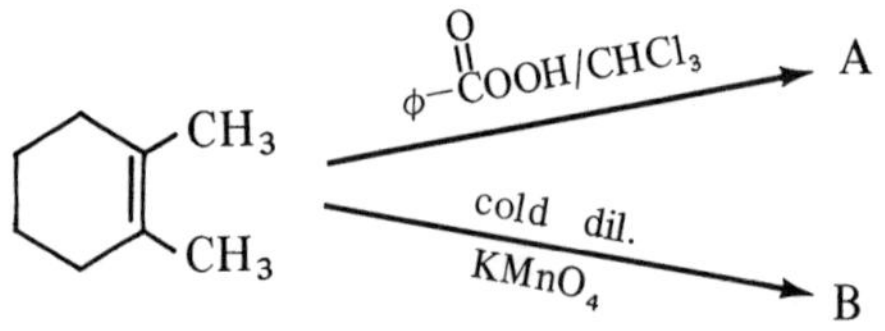

1,2-dimethylcyclohexene + φ-COOH/CHCl$_3$ → A

1,2-dimethylcyclohexene + cold dil. KMnO$_4$ → B

**Chapter 7
Alkenes**

**54. What is X? What is the mechanism?

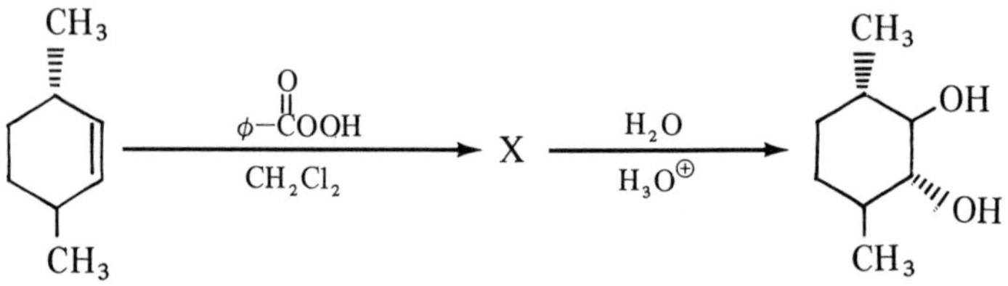

(1R)-1-methyl-(4R)-4-methylcyclohex-2-ene with CH$_3$ (wedge up) and CH$_3$ (wedge down) → [φ-COOH, CH$_2$Cl$_2$] → X → [H$_2$O, H$_3$O$^{\oplus}$] → cyclohexane with CH$_3$, OH, OH, CH$_3$ substituents (trans-diol product)

**Chapter 7
Alkenes**

*55. What is the structure of X, C$_{10}$H$_{14}$?

$(C_{10}H_{14})$ X $\xrightarrow[\text{2. Zn, H}_2\text{O}]{\text{1. O}_3}$ 1 mol of CH$_2$O and 1 mol of

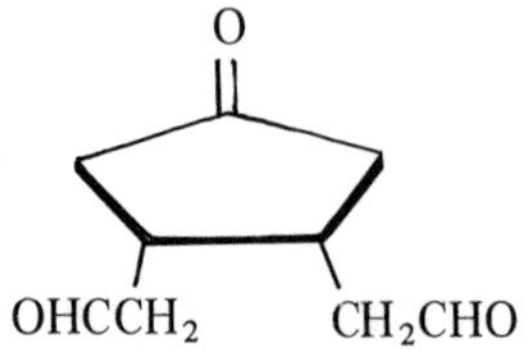

cyclopentanone with OHCCH$_2$– and –CH$_2$CHO substituents at the 2,5-positions

What is the name of the intermediate in this reaction, which is not isolated? Why is it not isolated?

**Chapter 7
Alkenes**

**56. Show a mechanism and the product.

$$CH_3CH_2CH_2CH=CH_2 \xrightarrow[\text{ROOR}]{\text{HBr } \Delta} X$$

**Chapter 7
Alkenes**

**57. An unknown compound was found leaking out of a cracked vial in the back of a freezer in a research lab. One of the people handling this material found it to be a powerful lachrymator. A mass spectrum showed a molecular weight of 161. Sodium fusion gave a positive test, and the precipitate was shown to contain bromide. Treatment of the compound with NaOEt in ethanol produced, among other things, a new compound containing no bromine, which when catalytically hydrogenated absorbed 2 moles of H_2 to give methylcyclopentane. The original unknown compound only absorbed 1 mole of H_2 upon catalytic hydrogenation. It also rapidly decolorized bromine. Ozonolysis of the unknown compound yielded a single keto–aldehyde, which gave a positive iodoform test. This keto–aldehyde also readily lost the elements of HBr to produce a compound with IR absorptions characteristic of an α,β-unsaturated aldehyde. The original compound rapidly gave a precipitate upon reaction with $AgNO_3$ in aqueous ethanol. Hydrogenation (1 mole) produced a compound that also gave a precipitate with $AgNO_3$, but much less rapidly. What is the structure of the unknown compound?

Chapters 5–7
Alkenes

**58. What is the product of the following reaction? Write a mechanism.

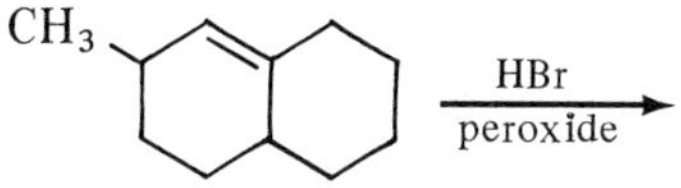

Chapter 7
Alkenes

*59. Show the major product and write a mechanism

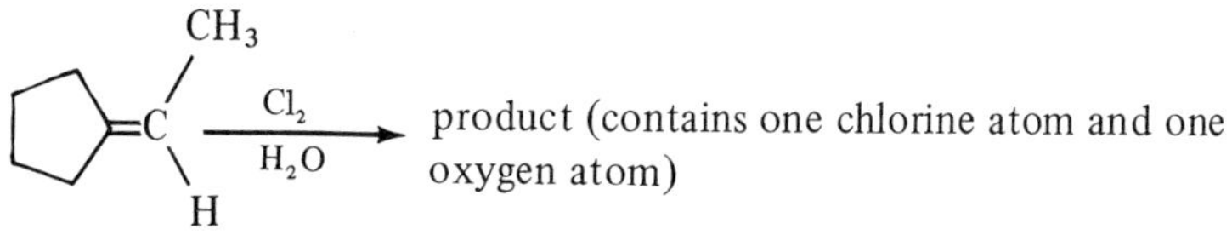

Chapters 7 and 8
Stereochemistry

**60. Describe in detail the products formed in the following reaction. Show how they are formed. Use three-dimensional pictorial representations.

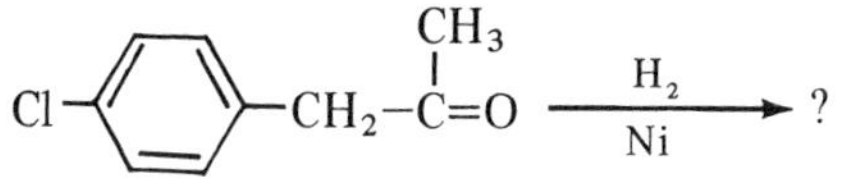

Chapter 8
Stereochemistry

32 PROBLEMS

*61. Consider the following Fischer projection (horizontal groups pointing out at you, vertical groups pointing back).

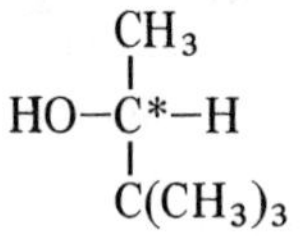

$$\begin{array}{c} CH_3 \\ | \\ HO-C^*-H \\ | \\ C(CH_3)_3 \end{array}$$

Show how you determine whether the stereogenic (chiral) carbon atom is (R) or (S).

**Chapter 8
Stereochemistry**

**62. Draw the structure of the product.

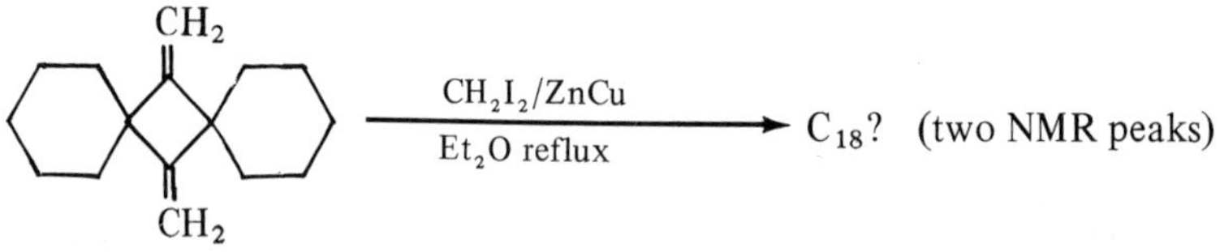

[Structure: spiro bicyclohexane-cyclobutane with two exocyclic =CH₂ groups] $\xrightarrow[\text{Et}_2\text{O reflux}]{\text{CH}_2\text{I}_2/\text{ZnCu}}$ C_{18}? (two NMR peaks)

**Chapter 8 and
Special Topic C
and Alkenes**

*63. Show the structure of the polymer that is formed by treating the following compound with a peroxide initiator as shown in the following equation.

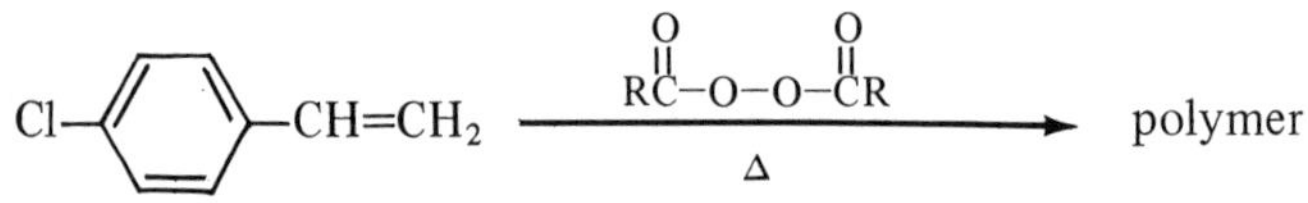

$$\text{Cl}-\underset{}{\underset{}{\bigcirc}}-CH=CH_2 \xrightarrow[\Delta]{\overset{O}{\overset{\|}{R C}}-O-O-\overset{O}{\overset{\|}{C R}}} \text{polymer}$$

Show a sequence of steps that results in the final polymeric product.

**Chapter 8 and
Special Topic B
Polymers**

*64. Show three major mechanisms of polymerization of an olefinic species. Indicate clearly the direction of electron flow in each case, and what R groups would facilitate the process in each case.

$R_2C=CR'_2 \longrightarrow$ polymer

**Chapter 8 and
Special Topic B
Polymers**

***65. Show the products and mechanisms of the following reactions:

cis-2-pentene + CH_2N_2 $\xrightarrow[\text{liq. phase}]{h\nu}$ one product

trans-2-pentene + CH_2N_2 $\xrightarrow[\substack{\text{gas phase with}\\\text{inert gas diluent}}]{h\nu}$ two products

Chapter 8 and
Special Topic C
Alkenes

**66. What are the products of the following reaction and what are the approximate yields of each?

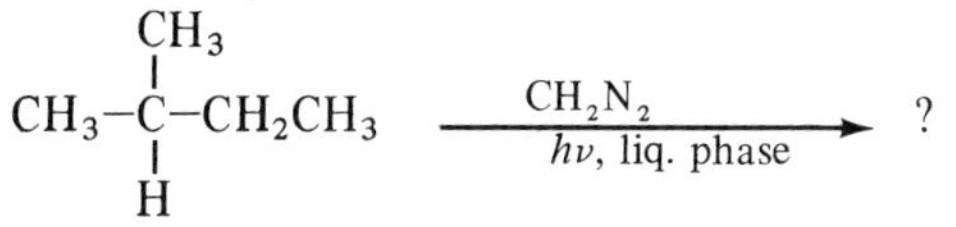

$$CH_3-\underset{\underset{H}{|}}{\overset{\overset{CH_3}{|}}{C}}-CH_2CH_3 \xrightarrow[h\nu,\text{ liq. phase}]{CH_2N_2} ?$$

Chapter 8 and
Special Topic C
Alkenes

***67. Show mechanisms for the following transformations, indicating the products at each step:

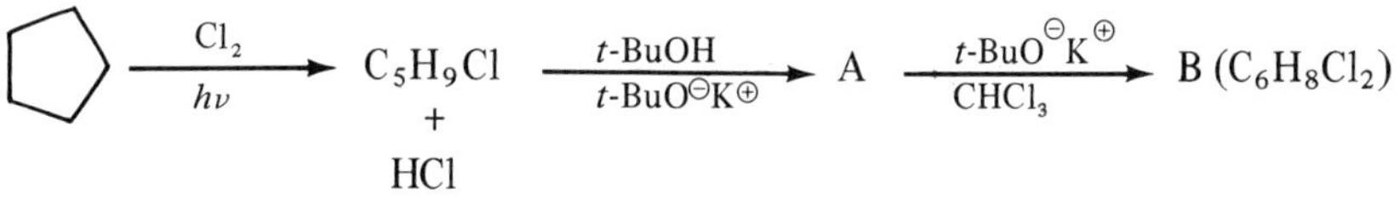

cyclopentane $\xrightarrow[h\nu]{Cl_2}$ C_5H_9Cl + HCl $\xrightarrow[\text{t-BuO}^\ominus K^\oplus]{t\text{-BuOH}}$ A $\xrightarrow[CHCl_3]{t\text{-BuO}^\ominus K^\oplus}$ B ($C_6H_8Cl_2$)

Chapters 4–6 and
Special Topic B
Alkenes

**68.

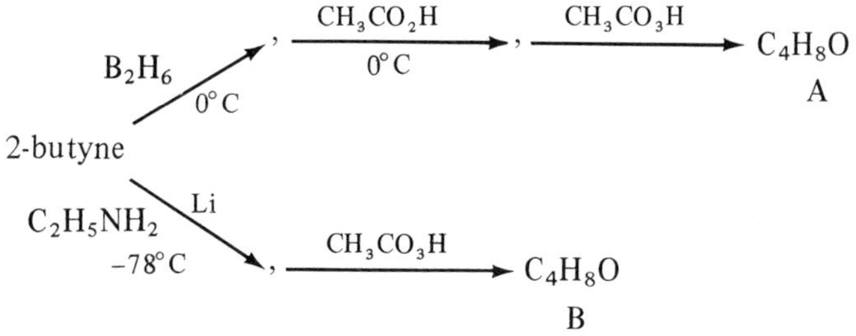

2-butyne $\xrightarrow[0°C]{B_2H_6}$, $\xrightarrow[0°C]{CH_3CO_2H}$, $\xrightarrow{CH_3CO_3H}$ C_4H_8O A

2-butyne $\xrightarrow[-78°C]{C_2H_5NH_2 \; \overset{Li}{}}$, $\xrightarrow{CH_3CO_3H}$ C_4H_8O B

What are the products A and B and what is the relationship between them?

Chapter 9
Alkynes

34 PROBLEMS

*69. Show the mechanism and product of the following reaction.

$$CH_3CH_2C\equiv C-H \xrightarrow[H_2O \;\; H^{\oplus}]{Hg^{2\oplus}} \;?$$

Chapter 9
Alkynes

**70. Using calcium carbide, toluene, and any inorganic reagents show how you would prepare 1-phenyl-*cis*-2-butene.

Chapter 9
Alkynes

*71. What is the major product that results from the addition of two equivalents of HCl to 1-pentyne and *why* is it the major product?

Chapter 9
Alkynes

***72. Fill in the missing reagents and/or products in the following sequence of reactions:

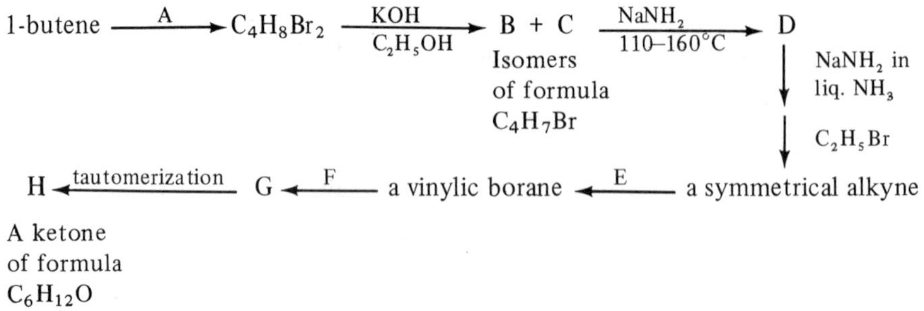

1-butene $\xrightarrow{A}$ $C_4H_8Br_2$ $\xrightarrow[C_2H_5OH]{KOH}$ B + C (Isomers of formula C_4H_7Br) $\xrightarrow[110-160°C]{NaNH_2}$ D $\xrightarrow{NaNH_2 \text{ in liq. } NH_3}$ $\xrightarrow{C_2H_5Br}$ a symmetrical alkyne $\xrightarrow{E}$ a vinylic borane $\xrightarrow{F}$ G $\xrightarrow{\text{tautomerization}}$ H

A ketone of formula $C_6H_{12}O$

Chapters 5–7 and 9
Alkynes

***73. A student tried to prepare an internal alkyne in the following way:

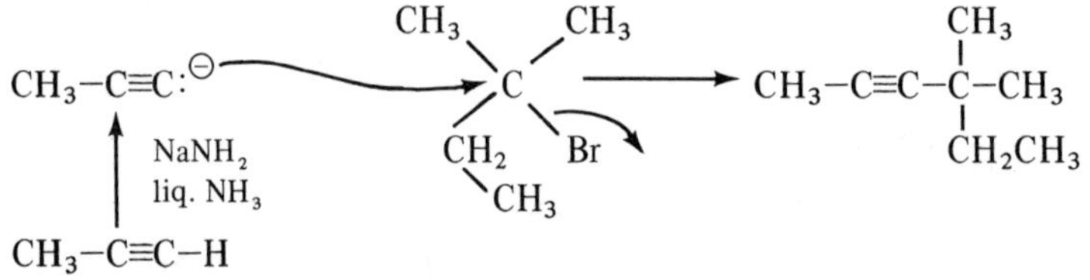

$$CH_3-C\equiv C{:}^{\ominus} + \underset{\underset{CH_3}{|}}{\overset{\overset{CH_3}{|}}{C}}\!\!\begin{array}{c}CH_3\\ \diagup\\ \diagdown \\ Br\end{array} \longrightarrow CH_3-C\equiv C-\underset{\underset{CH_2CH_3}{|}}{\overset{\overset{CH_3}{|}}{C}}-CH_3$$

$CH_3-C\equiv C-H \xrightarrow{\underset{\text{liq. } NH_3}{NaNH_2}}$

Upon work-up of the reaction mixture, the original alkyne was isolated, along with *two* new unsaturated compounds, neither of which was the desired alkyne. Explain.

Chapters 5, 6, and 9
Alkynes

PROBLEMS

**74. Write a detailed mechanism for the addition of HBr to butadiene at high temperature. Explain the ratio of products formed.

Chapter 10
Conjugated Systems

**75. What are the products obtained by heating cyclohexadiene with the products resulting from heating 2-bromosuccinic acid dimethyl ester with a strong, hindered tertiary amine?

alkoxide in answer

Chapters 5, 6, and 10
Conjugated Systems

***76. The energies of the molecular orbitals of conjugated cyclic systems can be graphically depicted by inscribing the geometric shape of the cycle (point down) inside a circle. The atoms at the vertices reflect the energy levels of the MO's. Benzene is shown here in this way. Those orbitals below the midline of the circle are bonding, those above it are antibonding, and those on the line are nonbonding. Benzene has no nonbonding MO's, but other systems do.

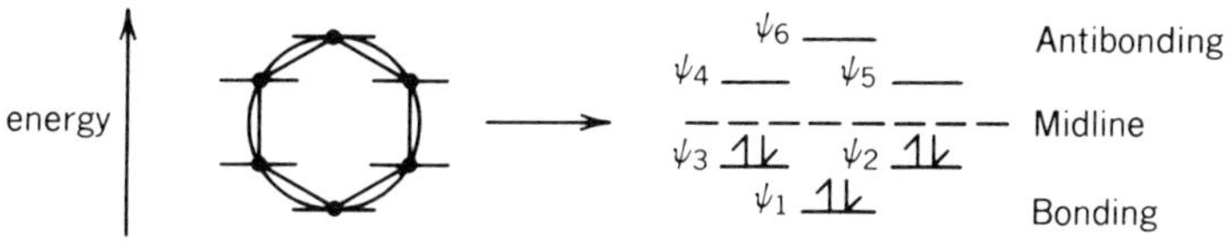

Using this method to elaborate MO's, answer the following questions:

(a) Is cyclobutadiene stable or not? Explain.

(b) Is cyclopentadiene acidic or not? Explain. (Consider its acidity compared to cyclopentane.)

(c) Is cyclooctatetraene aromatic? Explain.

Chapter 11
Aromatic Compounds

**77. Starting with 2-phenyl-2-butene and the reagents D_2O, HBr, magnesium metal, and an organic peroxide, show how you could prepare 2-deuterio-4-phenylbutane. Show a mechanism for the first step in your synthetic sequence.

Chapters 7 and 11
Aromatic Compounds

36 PROBLEMS

***78. Show all the products and intermediates. Show a mechanism for the last step in the following sequence:

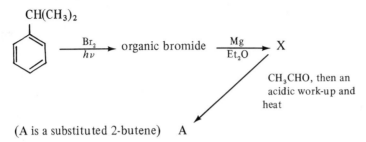

(A is a substituted 2-butene)

**Chapters 6 and 11
Conjugated Systems**

*79. What is the mechanism and major product of the following reaction?

CH$_3$—C$_6$H$_5$ $\xrightarrow[\text{AlCl}_3]{\text{CH}_3\text{CHCH}_3, \text{Cl}}$?

**Chapter 12
Aromatic Compounds**

*80. Which of the following represents the best route to *m*-bromobenzoic acid from benzene.

(a) C$_6$H$_6$ $\xrightarrow[\text{AlCl}_3]{\text{CH}_3\text{Cl}}$ $\xrightarrow[\text{Fe}]{\text{Br}_2}$ $\xrightarrow{\text{KMnO}_4}$

(b) C$_6$H$_6$ $\xrightarrow[\text{Fe}]{\text{Br}_2}$ $\xrightarrow[\text{AlCl}_3]{\text{CH}_3\text{Cl}}$ $\xrightarrow{\text{KMnO}_4}$

(c) C$_6$H$_6$ $\xrightarrow[\text{AlCl}_3]{\text{CH}_3\text{Cl}}$ $\xrightarrow{\text{KMnO}_4}$ $\xrightarrow[\text{Fe}]{\text{Br}_2}$

**Chapter 12
Aromatic Compounds**

*81. A student attempted to run a Friedel–Crafts methylation on 4-chloroaniline using methyl chloride with an aluminum chloride catalyst. Upon mixing the reagents a precipitate formed and the reaction did not proceed. What happened? Explain in detail.

**Chapter 12
Aromatic Compounds**

*82. Draw the structure of the sigma complex intermediate formed by attack at the para position of nitrobenzene when this substrate is nitrated with a mixture of nitric acid and fuming sulfuric acid. Explain why this intermediate is favored or not favored compared to the intermediate formed by attack at the meta position. What will be the favored product?

**Chapter 12
Aromatic Compounds**

**83. Why do the following reactions yield more para than ortho substitution product?

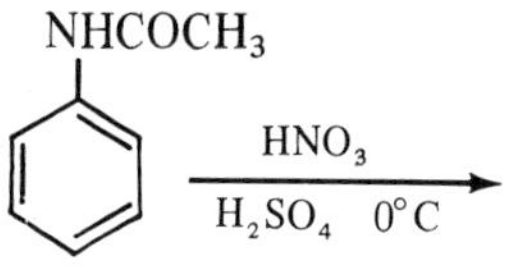

NHCOCH$_3$–C$_6$H$_5$ $\xrightarrow[\text{H}_2\text{SO}_4 \quad 0°\text{C}]{\text{HNO}_3}$ Nitroacetanilide products

19% ortho   79% para   2% meta

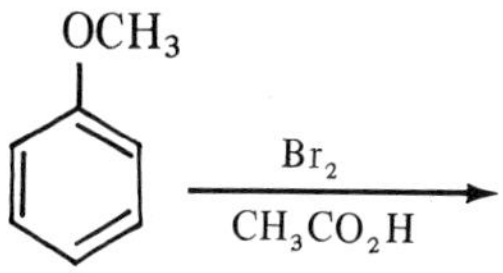

OCH$_3$–C$_6$H$_5$ $\xrightarrow[\text{CH}_3\text{CO}_2\text{H}]{\text{Br}_2}$ Bromoanisole products

4% ortho   96% para   0% meta

**Chapter 12
Aromatic Compounds**

**84. What do you think would be the major products obtained by monobromination of 2-chloroaniline with bromine in the presence of a small amount of FeBr$_3$? Explain your answer.

**Chapters 1 and 12
Aromatic Compounds**

*85. A student tried to prepare isobutylbenzene by reacting (CH$_3$)$_2$CHCH$_2$Cl with benzene in the presence of an aluminum chloride catalyst. The product obtained was a mixture that had an NMR spectrum showing absorptions in the aliphatic region not characteristic of those expected for the product. What might have happened?

**Chapter 12
Aromatic Compounds**

***86. How could you make 1,3,5-trinitrobenzene in three steps starting from toluene and inorganic reagents? Explain why the last step (a decarboxylation . . . this is a hint to help you design your synthesis) is so facile.

**Chapter 12
Aromatic Compounds**

38 PROBLEMS

**87. An attempt was made to add methyl magnesium bromide to the carbonyl function of the following compound:

$$CH_3\overset{\overset{O}{\|}}{C}CH(CN)_2 \quad \xrightarrow{CH_3MgBr}{Et_2O}$$

Small amounts of a gaseous product were observed bubbling from the reaction mixture. Explain.

Chapter 14
Organic Halides

***88. A student attempted the preparation of 1-cyanobutane by rapidly stirring a heated mixture of 1-bromobutane in methylene chloride with an aqueous solution of sodium cyanide. The reaction never worked at all unless a specific bottle of methylene chloride was used, and even then the yields were not high. This bottle had the words "crude—to be distilled later" written in pencil underneath the label name of methylene chloride. The student never bothered to take note of this until he/she ran out of this particular source of the solvent. At this point he/she inquired about the source of the crude solvent. It apparently had been obtained from the collection flask of a rotary evaporator that was used to remove methylene chloride from a mixture of triethylamine and various deuterated triethylamines. Explain why the crude solvent gave better yields than pure methylene chloride. (*Hint:* The student who had used the rotary evaporator to separate amines from methylene chloride did not continue to separate them this way; he/she found that careful fractional distillation was much more appropriate.)

Chapter 14
Organic Halides

**89.

[Reaction: 2-chloro-3,5-dinitrophenethyl alcohol + CH$_3$OH / CH$_3$OK → dinitroanisole product with (CH$_2$)$_2$OH and OCH$_3$ groups + X]

The preceding reaction was carried out to prepare the dinitroanisole indicated. In addition to this product, another material, X, was also isolated that contained no chlorine and that had a molecular weight the same as the starting material minus HCl. Its NMR spectrum was similar to that of the expected product, except that the methylene (upfield) region was more complicated. In addition, no O–H stretching band was observed in the IR spectrum of X. What is X and how is it formed?

Chapter 14
Organic Halides

90. Show two methods for preparing 1-phenylethanol starting with benzene. Use chloroethane in one method and ethanoyl chloride in the other.

Chapters 5, 12, and 15
Alcohols

*91. Which compound reacts most rapidly with sodium metal?

(a) Cyclohexanol

(b) Ethanol

(c) $CH_3CH_2OCH_2CH_3$

(d) 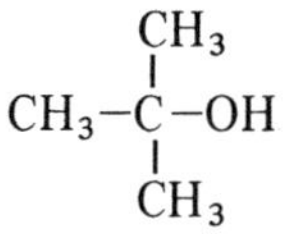 $CH_3-\underset{\underset{CH_3}{|}}{\overset{\overset{CH_3}{|}}{C}}-OH$

Chapter 15
Alcohols, Phenols, and Ethers

*92. Synthesize 3-methyl-1-butanol starting with 2-propanol and any necessary reagents.

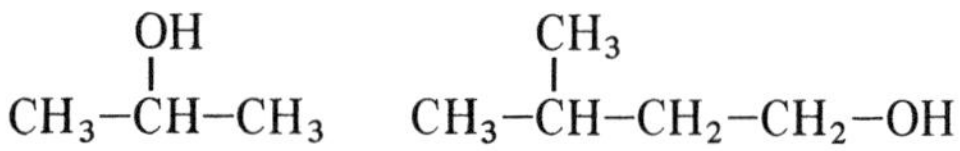

$CH_3-\underset{\underset{}{}}{\overset{\overset{OH}{|}}{CH}}-CH_3$ $CH_3-\underset{\underset{}{}}{\overset{\overset{CH_3}{|}}{CH}}-CH_2-CH_2-OH$

2-Propanol   3-Methyl-1-butanol

Chapter 15
Alcohols

*93. Starting with acetaldehyde as the only carbon source, show how to prepare ethyl acetate.

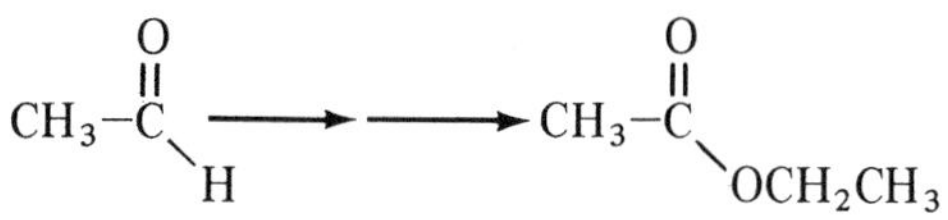

$CH_3-\overset{O}{\underset{}{\overset{\|}{C}}}-H$ ⟶ $CH_3-\overset{O}{\underset{}{\overset{\|}{C}}}-OCH_2CH_3$

Acetaldehyde   Ethyl acetate

Chapter 15
Alcohols and Esters

40 PROBLEMS

*94. Which of the following is the least stable carbanion?

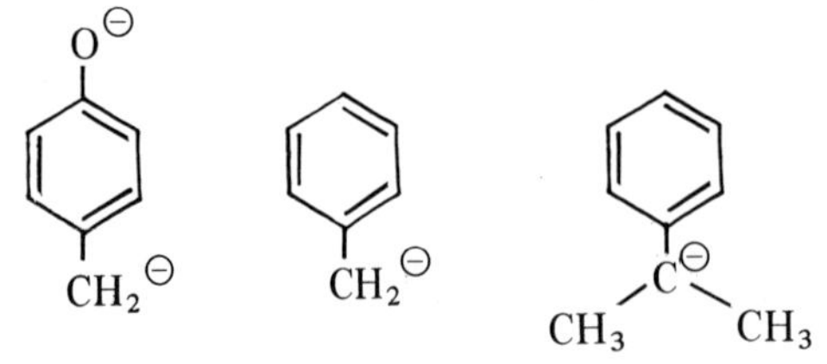

A   B   C

Chapters 2 and 15
Alcohols

**95. An attempt to carry out the following reaction failed.

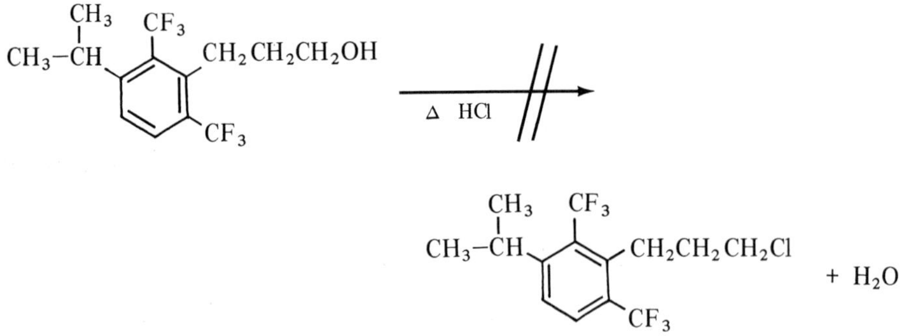

A small piece of zinc added to the reaction mixture resulted in a good yield of the desired product. Explain. Show a mechanism for the reaction.

Chapter 15
Alcohols

***96. Consider the following reaction:

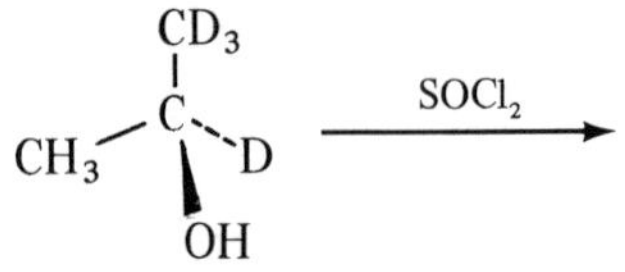

→ the major product is one with retained configuration

If the chlorination is carried out with a small amount of triethylammonium chloride present, a mixture of retained and inverted product can be obtained. Explain. Show mechanisms for these transformations.

Chapter 15
Alcohols

*97. What are the products of the reaction of isopropylmagnesium bromide with the following compounds? Assume an acid hydrolysis occurs in the work-up.

(a)  (CH₃)₂CH—MgBr + HCHO ⟶

(b) 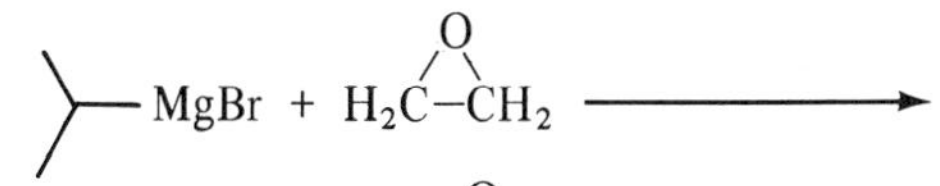 (CH₃)₂CH—MgBr + H₂C—CH₂ (epoxide, O) ⟶

(c) 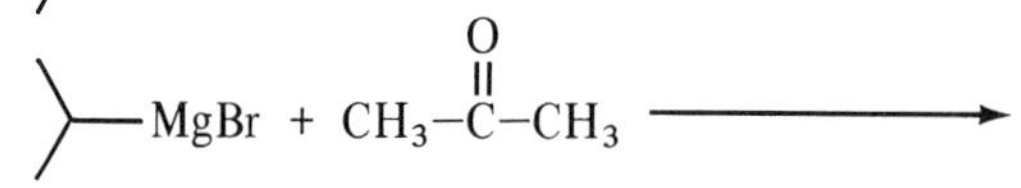 (CH₃)₂CH—MgBr + CH₃—C(=O)—CH₃ ⟶

(d) 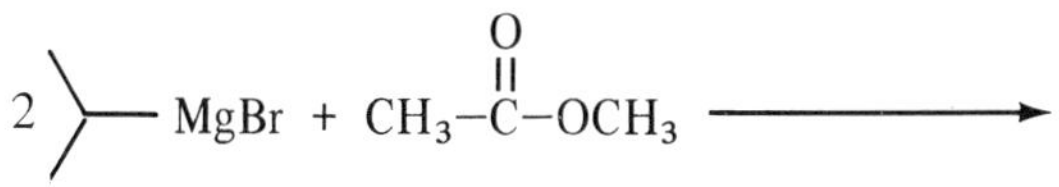 2 (CH₃)₂CH—MgBr + CH₃—C(=O)—OCH₃ ⟶

**Chapter 15
Grignard Reaction**

**98. What are the products? What are the mechanisms?

(1) 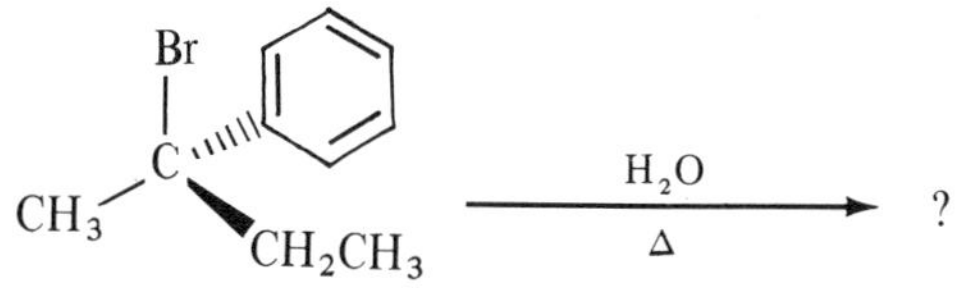 CH₃—C(Br)(Ph)(CH₂CH₃) $\xrightarrow[\Delta]{H_2O}$?

(2)  CH₃—C(Br)(H)(D) $\xrightarrow[NaI]{CH_3COCH_3}$?

**Chapters 5, 8, and
Special Topic H
Nucleophilic Displacements**

*99. Which of the following molecules will ionize most rapidly under polar conditions? Why? Which ionizes least rapidly? Why?

$$CH_3CH_2\underset{Br}{CH}CH_2CH_3 \qquad CH_3-O-\underset{Br}{CH}CH_2CH_3 \qquad CH_3CH_2\underset{Br}{CH}CF_2CF_3$$

A B C

**Chapters 1, 5, and Special Topic H
Alcohols**

42 PROBLEMS

***100.** In an effort to prepare methyl-*t*-butyl ether an introductory organic chemistry student decided on the Williamson synthesis using the following scheme:

$$CH_3OH \xrightarrow{Na} CH_3O^{\ominus}Na^{\oplus} \xrightarrow{(CH_3)_3CBr} (CH_3)_3COCH_3$$

The student was unsuccessful in obtaining the desired product. Why? What product would you expect?

Chapter 15
Alcohols

**101. What are the products of the following reaction?

Ph—O—C(CH$_3$)$_3$ $\xrightarrow[\text{HBr}]{48\%}$?

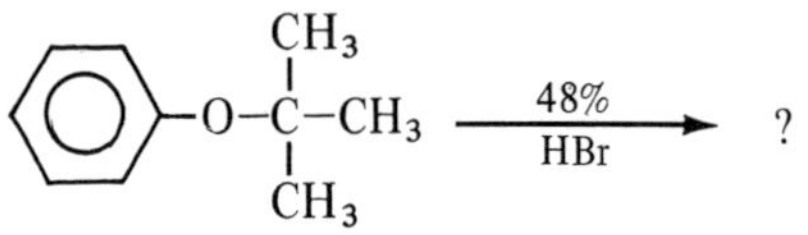

Write a reasonable mechanism.

Chapter 15
Alcohols

**102. Show the product and suggest a mechanism for the following reaction:

1-hydroxy-1,2,2-trimethyl... (cyclopentane with CH$_3$, OH, CH$_3$, CH$_3$, CH$_3$ substituents) $\xrightarrow[\text{Et}_2O]{CH_3Li}$ $\xrightarrow{p\text{-toluenesulfonyl-chloride}}$ C$_{16}$H$_{24}$O$_3$S

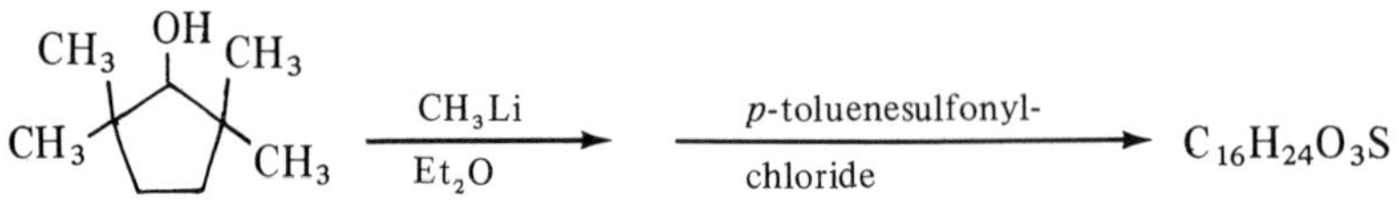

Chapter 15
Alcohols

*103. Use the Williamson ether synthesis to prepare anisole:

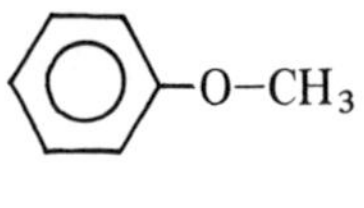

Ph—O—CH$_3$

Anisole

Chapter 15
Alcohols

PROBLEMS

*104. Suggest a mechanism for the following conversion:

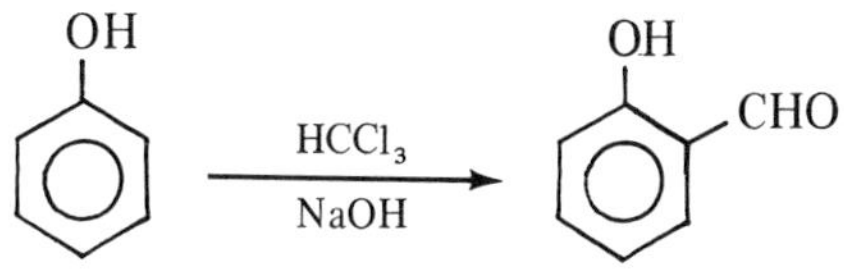

phenol + HCCl$_3$ / NaOH → salicylaldehyde (2-hydroxybenzaldehyde)

Chapter 15
Alcohols

***105. Write a mechanism for the following:

1,1'-bicyclopentyl-1,1'-diol + H$^\oplus$ → spiro[4.5]decan-6-one

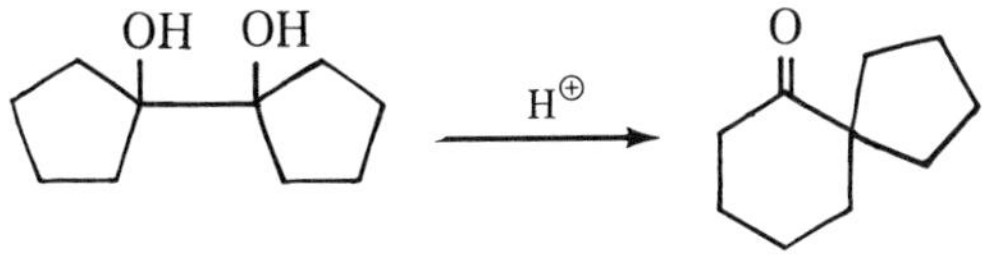

Chapter 15
Alcohols

***106. Account for the following transformation by writing an acceptable mechanism:

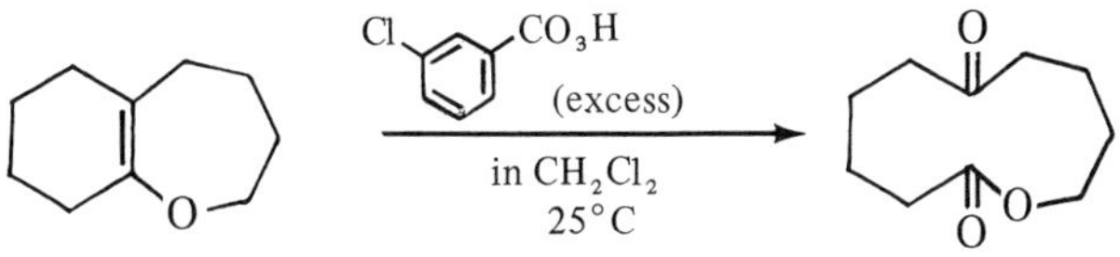

2,3,4,5-tetrahydro-1-benzoxepine + m-chloroperbenzoic acid (excess) in CH$_2$Cl$_2$ at 25°C → dioxo lactone product

Chapter 15
Ethers

***107. Outline a mechanistic route for the formation of 1,2-dimethylcyclopentene from 2-cyclobutyl-2-propanol in sulfuric acid.

Chapter 15
Alcohols

**108. Which is more acidic? Why?

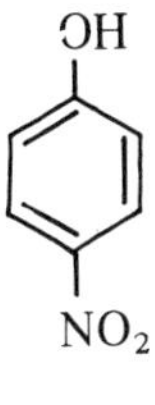

A: 4-nitrophenol
B: 3,5-di-*t*-butyl-4-nitrophenol

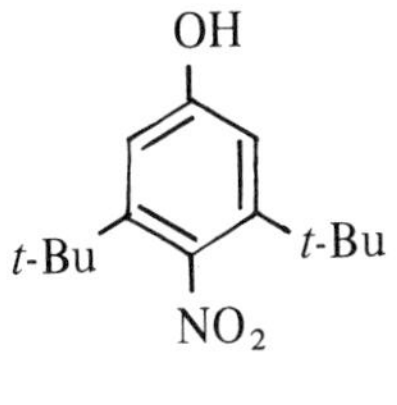

Chapter 15
Phenols

44 PROBLEMS

***109. Write a mechanism that accounts for the following:

$$(CH_3)_2CHOH + H_2CrO_4 \longrightarrow CH_3-\underset{\underset{O}{\|}}{C}-CH_3$$

Chapters 15 and 16
Alcohols and Ketones

*110. Suggest a mechanism for the following reaction:

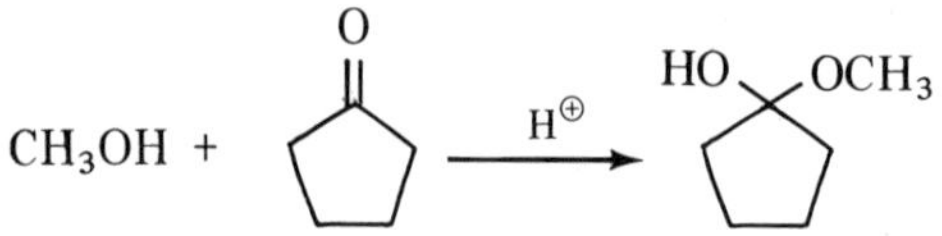

$$CH_3OH + \text{cyclopentanone} \xrightarrow{H^{\oplus}} \text{1-hydroxy-1-methoxycyclopentane}$$

Chapter 16
Aldehydes and Ketones

*111. Reaction of cyclohexanone with diazomethane produces cycloheptanone:

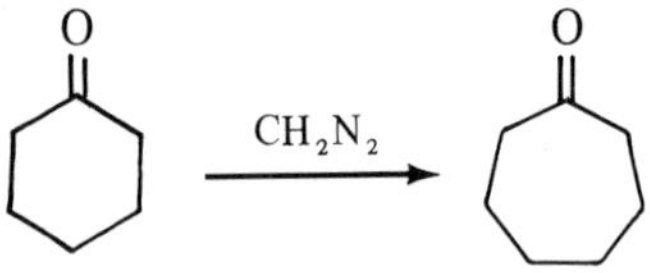

$$\text{cyclohexanone} \xrightarrow{CH_2N_2} \text{cycloheptanone}$$

Suggest a mechanism for this reaction.

Chapter 16
Aldehydes and Ketones

*112. Provide structures for Compounds A and B. Suggest a mechanism for the formation of Compound A.

$$\text{butanone} \xrightarrow{HCN} \underset{C_5H_9NO}{A} \xrightarrow[H_2O]{HCl} \underset{C_5H_{10}O_3}{B}$$

Chapter 16
Aldehydes and Ketones

*113. Provide a mechanism for the following:

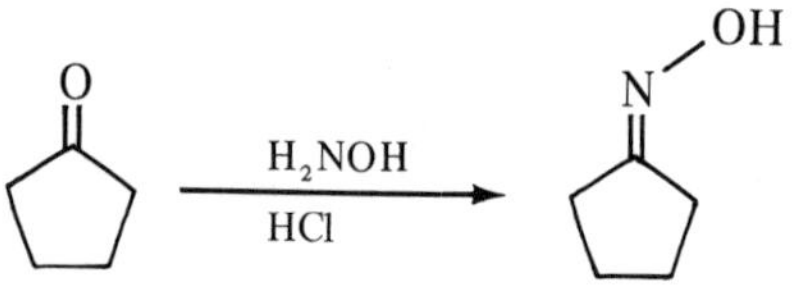

$$\text{cyclopentanone} \xrightarrow[HCl]{H_2NOH} \text{cyclopentanone oxime}$$

Chapter 16
Aldehydes and Ketones

*114. Draw structures and give the general name for the products in the following reactions.

(a) 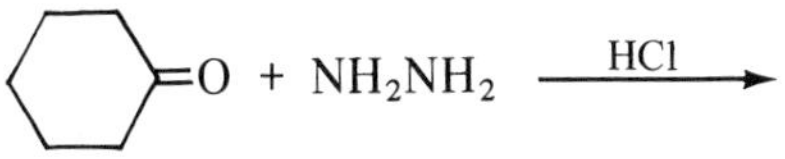 cyclohexanone + NH_2NH_2 $\xrightarrow{HCl}$

(b) 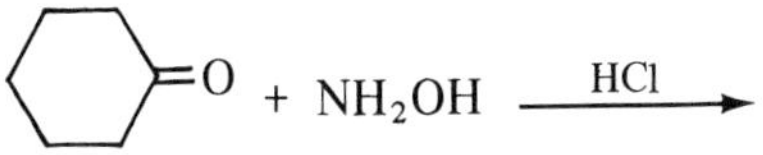 cyclohexanone + NH_2OH $\xrightarrow{HCl}$

(c) 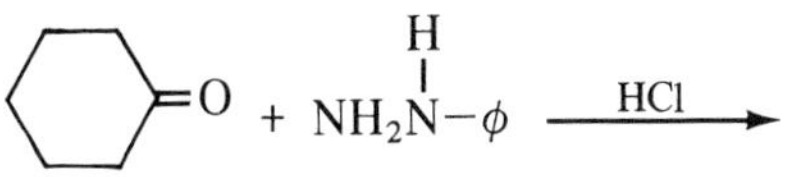 cyclohexanone + $NH_2N(H)-\phi$ $\xrightarrow{HCl}$

(d) 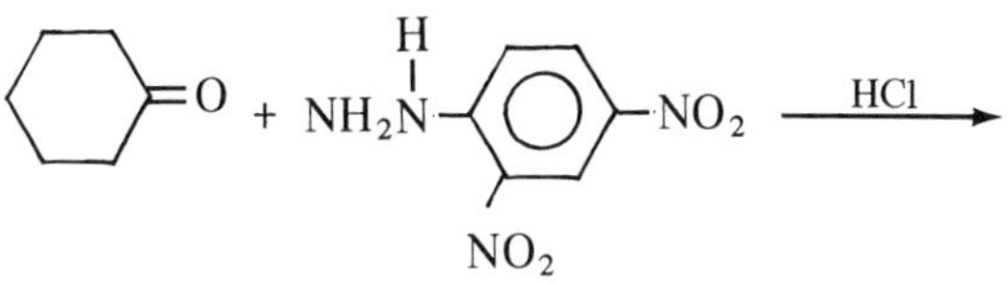 cyclohexanone + $NH_2N(H)$–(2,4-dinitrophenyl) $\xrightarrow{HCl}$

**Chapter 16
Ketones**

**115. Provide a mechanism for the following:

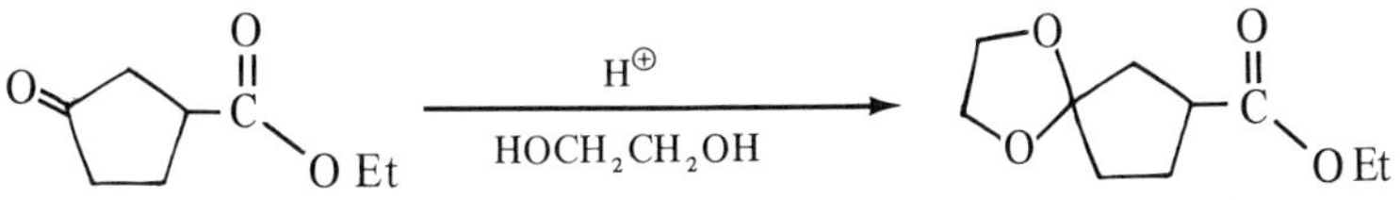

3-oxocyclopentanecarboxylic acid ethyl ester $\xrightarrow[\text{HOCH}_2\text{CH}_2\text{OH}]{H^{\oplus}}$ ketal product with –COOEt

**Chapter 16
Aldehydes and Ketones**

***116. Suggest a mechanism for the following:

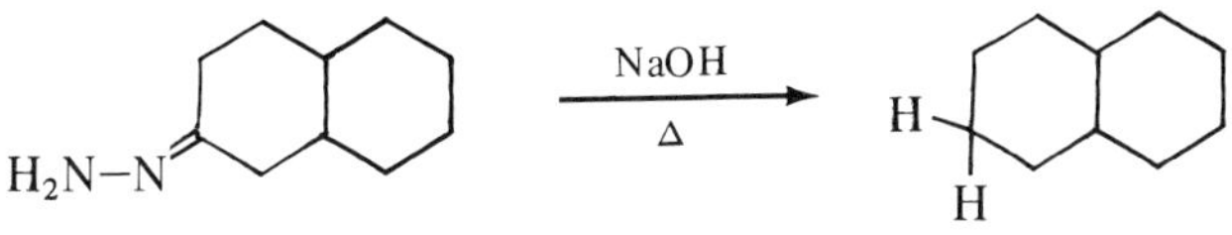

$H_2N-N=$(decalin-2-ylidene) $\xrightarrow[\Delta]{NaOH}$ decalin (with H's shown)

**Chapter 16
Aldehydes and Ketones**

117. What is the structure of the final product in the following reaction sequence?

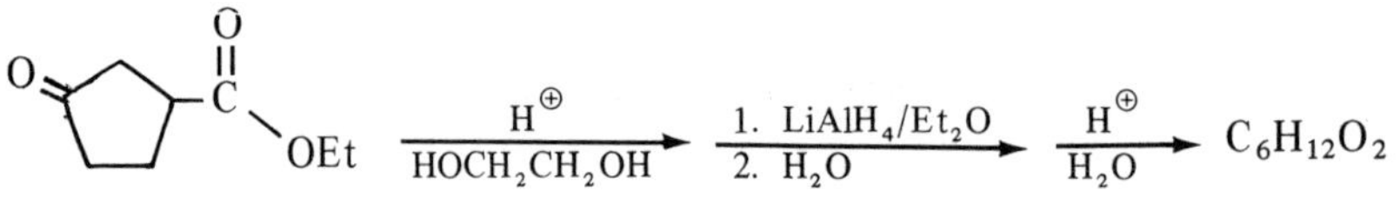

3-oxocyclopentanecarboxylic acid ethyl ester $\xrightarrow[\text{HOCH}_2\text{CH}_2\text{OH}]{\text{H}^{\oplus}}$ $\xrightarrow[\text{2. H}_2\text{O}]{\text{1. LiAlH}_4/\text{Et}_2\text{O}}$ $\xrightarrow[\text{H}_2\text{O}]{\text{H}^{\oplus}}$ $C_6H_{12}O_2$

An infrared spectrum of the product shows absorption at 1750 and 3500 cm^{-1}.

Chapter 16
Aldehydes and Ketones

118. Explain the following:

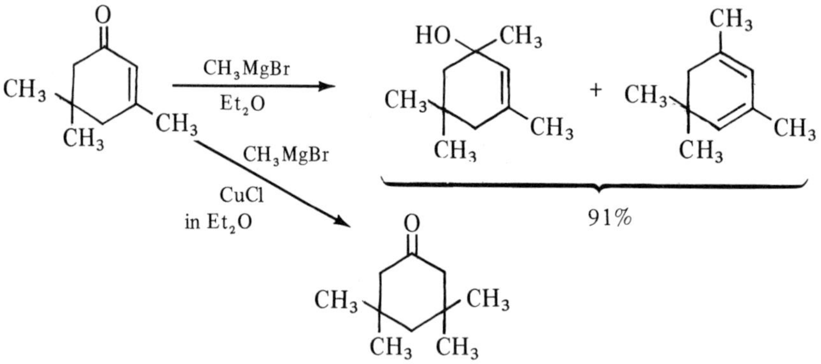

3,5,5-trimethylcyclohex-2-enone $\xrightarrow[\text{Et}_2\text{O}]{\text{CH}_3\text{MgBr}}$ 1-hydroxy-1,3,5,5-tetramethylcyclohex-2-ene + 1,3,5-trimethylbenzene (91%)

$\xrightarrow[\text{in Et}_2\text{O}]{\text{CH}_3\text{MgBr, CuCl}}$ 3,3,5,5-tetramethylcyclohexanone

Chapter 16
Aldehydes and Ketones

119. Supply the structures for **II–V**.

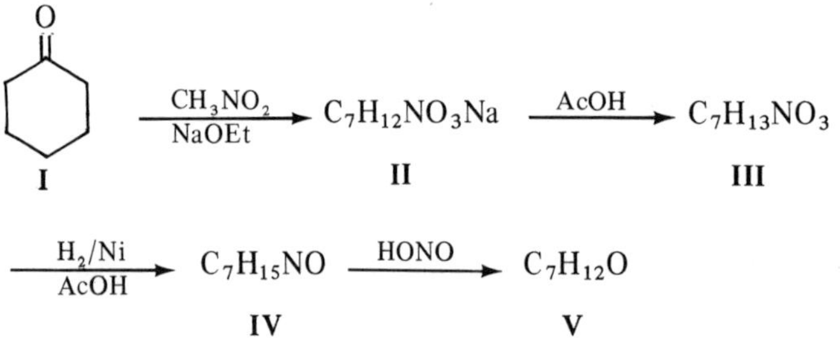

cyclohexanone (**I**) $\xrightarrow[\text{NaOEt}]{\text{CH}_3\text{NO}_2}$ $C_7H_{12}NO_3Na$ (**II**) $\xrightarrow{\text{AcOH}}$ $C_7H_{13}NO_3$ (**III**)

$\xrightarrow[\text{AcOH}]{\text{H}_2/\text{Ni}}$ $C_7H_{15}NO$ (**IV**) $\xrightarrow{\text{HONO}}$ $C_7H_{12}O$ (**V**)

Chapter 16
Ketones

***120. Write a mechanism for the following reaction:

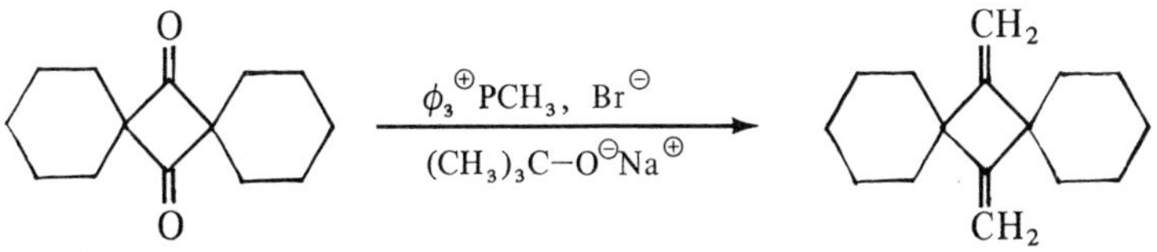

[diketone spiro bis-cyclohexane structure] $\xrightarrow[(CH_3)_3C-O^\ominus Na^\oplus]{\phi_3^\oplus PCH_3, \ Br^\ominus}$ [bis-methylene product]

Chapter 16
Ketones

***121. Provide a structure for the product and suggest a mechanism.

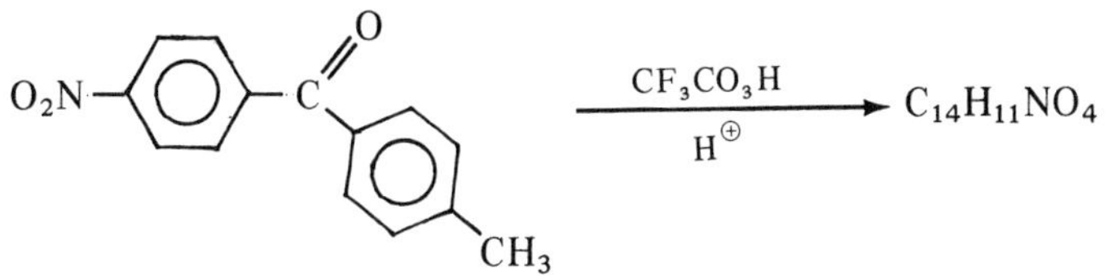

$O_2N-\text{C}_6H_4-\overset{O}{\underset{}{C}}-\text{C}_6H_4-CH_3 \xrightarrow[H^\oplus]{CF_3CO_3H} C_{14}H_{11}NO_4$

Chapter 16
Ketones

***122. Suggest a mechanism for the following transformation:

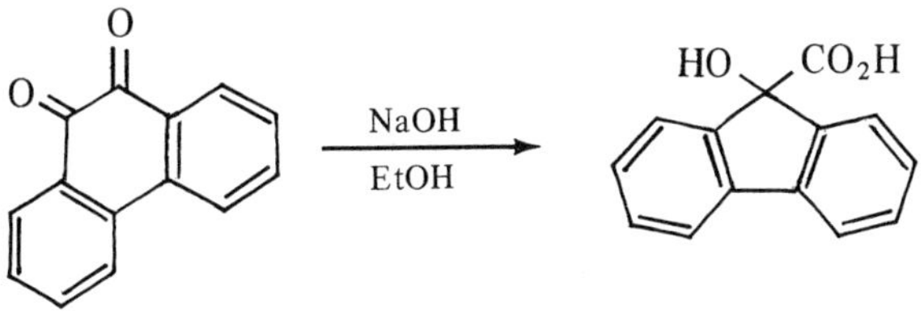

[phenanthrenequinone] $\xrightarrow[EtOH]{NaOH}$ [9-hydroxy-9-carboxyfluorene]

Chapter 16
Ketones

**123. What reagents are needed to synthesize the following compound?

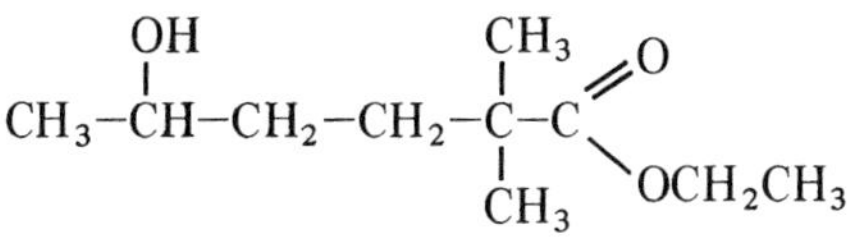

$$CH_3-\underset{\underset{OH}{|}}{CH}-CH_2-CH_2-\underset{\underset{CH_3}{|}}{\overset{\overset{CH_3}{|}}{C}}-\underset{}{\overset{O}{\overset{\|}{C}}}-OCH_2CH_3$$

Chapter 16
Aldehydes and Ketones

48 PROBLEMS

***124.** Draw structures for the products.

(a) 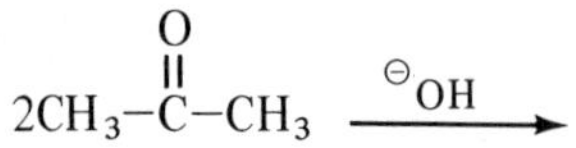 $2 \, CH_3-\overset{\overset{\displaystyle O}{\|}}{C}-CH_3 \xrightarrow{\ ^{\ominus}OH\ }$

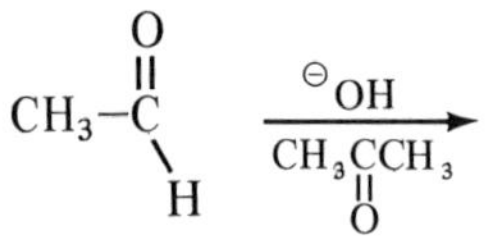

(b) $CH_3-\overset{\overset{\displaystyle O}{\|}}{C}\diagdown_H \xrightarrow[CH_3CCH_3 \atop \| \atop O]{\ ^{\ominus}OH\ }$

(c) $2 \, CH_3-\overset{\overset{\displaystyle O}{\|}}{C}\diagdown_H \xrightarrow{\ ^{\ominus}OH\ }$

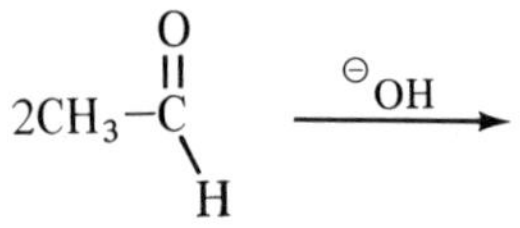

**Chapter 17
Aldehydes and Ketones**

***125.** Suggest a mechanism for the following transformation:

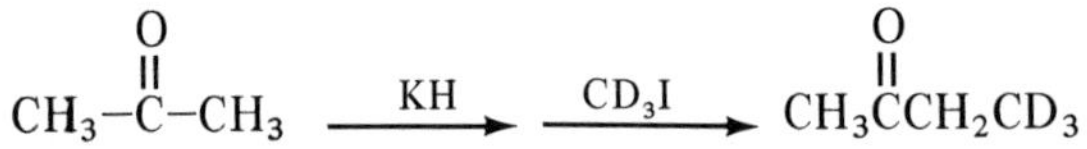

$CH_3-\overset{\overset{\displaystyle O}{\|}}{C}-CH_3 \xrightarrow{KH} \xrightarrow{CD_3I} CH_3\overset{\overset{\displaystyle O}{\|}}{C}CH_2CD_3$

**Chapter 17
Aldehydes and Ketones**

****126.** Write a reasonable mechanism for the following:

$\phi-\overset{\overset{\displaystyle O}{\|}}{C}\diagdown_H \xrightarrow{NaCN} \phi-\overset{\overset{\displaystyle O}{\|}}{C}-\overset{\overset{\displaystyle OH}{|}}{\underset{\underset{\displaystyle H}{|}}{C}}-\phi$

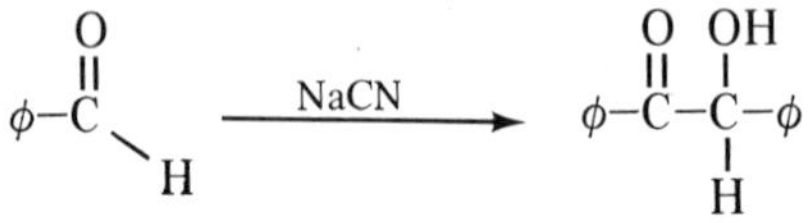

**Chapter 17
Ketones and Aldehydes**

****127.** Suggest a mechanism for the following:

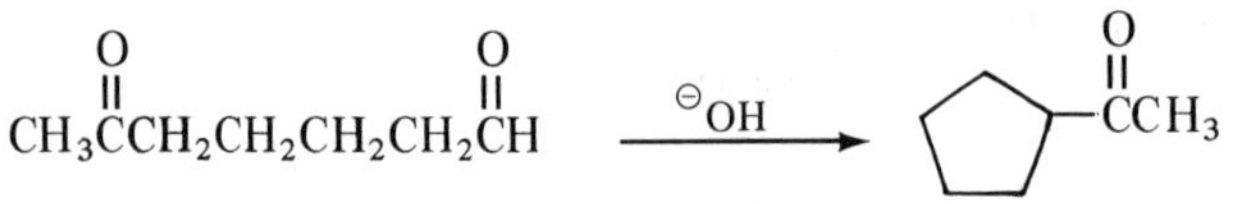

$CH_3\overset{\overset{\displaystyle O}{\|}}{C}CH_2CH_2CH_2CH_2\overset{\overset{\displaystyle O}{\|}}{C}H \xrightarrow{\ ^{\ominus}OH\ }$ [cyclopentane ring]$-\overset{\overset{\displaystyle O}{\|}}{C}CH_3$

**Chapter 17
Aldehydes and Ketones**

128. Account for the formation of acetone from the basic hydrolysis of a 5% solution of

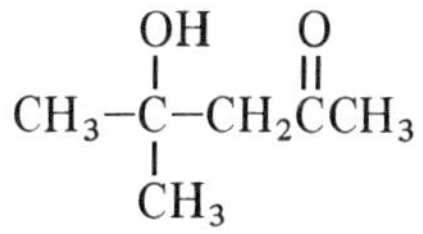

$$CH_3-\underset{\underset{CH_3}{|}}{\overset{\overset{OH}{|}}{C}}-CH_2\overset{\overset{O}{\|}}{C}CH_3$$

Chapter 17
Aldehydes and Ketones

129. Present a mechanism that illustrates how formaldehyde serves as a hydride source in this typical Cannizzaro reaction.

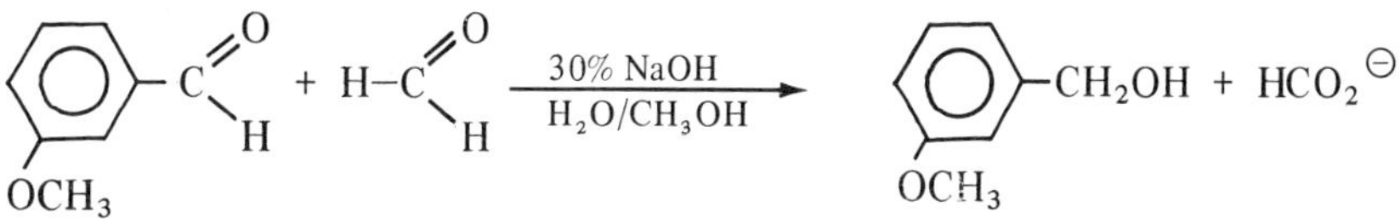

m-methoxybenzaldehyde + H–CHO $\xrightarrow[\text{H}_2\text{O}/\text{CH}_3\text{OH}]{30\% \text{ NaOH}}$ *m*-methoxybenzyl alcohol + $HCO_2^{\ominus}$

Chapter 17
Aldehydes and Ketones

130. Provide a mechanism for the following transformation:

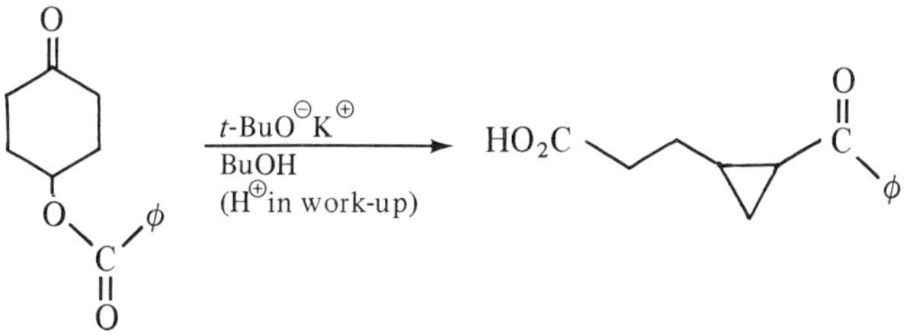

4-oxocyclohexyl benzoate $\xrightarrow[\text{BuOH} \atop (\text{H}^{\oplus} \text{ in work-up})]{t\text{-BuO}^{\ominus}\text{K}^{\oplus}}$ $HO_2C\text{-CH}_2\text{CH}_2\text{-cyclopropyl-C(O)}\phi$

Chapter 17
Aldehydes and Ketones

131. Write a mechanism for the following:

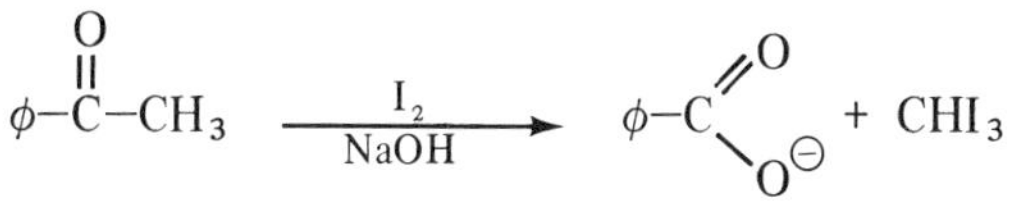

$\phi-\overset{\overset{O}{\|}}{C}-CH_3 \xrightarrow[\text{NaOH}]{I_2} \phi-C\overset{\overset{O}{\diagup}}{\diagdown}_{O^{\ominus}} + CHI_3$

Chapter 17
Aldehydes and Ketones

***132.** The perfume constituent *cis*-jasmone is prepared by reacting the following compound with aqueous sodium hydroxide. Suggest a structure for the product that produces the following spectrophotometric data:

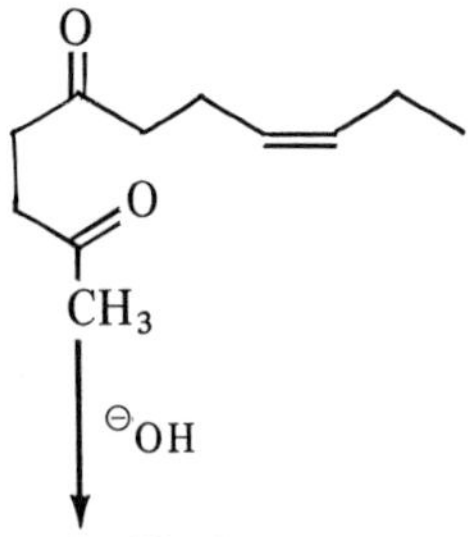

[Structure: a chain with C=O, CH₂CH₂, C(=O)CH₃, and a cis CH=CHCH₂CH₃ group; with ⁻OH arrow; C₁₁H₁₆O (Jasmone)]

UV: Max = 235 (12,000)

IR: 3430 cm⁻¹ (w) 2949 cm⁻¹ (m)
    3003 cm⁻¹ (w) 1702 cm⁻¹ (s)
    1652 cm⁻¹ (s)

NMR: δ 5.47 ppm (2H)
     3.03 ppm (2H)
     2.14 ppm (3H) (a singlet)
     1.02 ppm (3H)

Chapter 17
Aldehydes and Ketones

***133.** Show a mechanism that is consistent with the labeling information that follows:

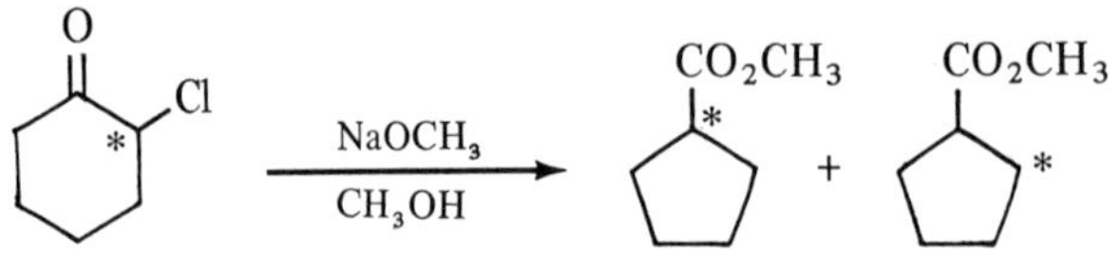

[2-chlorocyclohexanone with labeled * carbon → NaOCH₃/CH₃OH → methyl cyclopentanecarboxylate with * on ring carbon bearing CO₂CH₃ (50%) + methyl cyclopentanecarboxylate with * on adjacent ring carbon (50%)]

Chapter 17
Aldehydes and Ketones

*134. Propanoic acid reacts with phosphorous and bromine to produce A, C₃H₅O₂Br. Compound A reacts with two equivalents of ammonia to produce B, C₃H₇NO₂, and ammonium bromide. Compound B is amphoteric. Propose structures for A and B.

$$CH_3CH_2CO_2H \xrightarrow[Br_2]{P} A \xrightarrow{2NH_3} B + NH_4Br$$

Chapter 18
Acids

*135. Draw the structures of the products.

(a) $CH_3-\overset{O}{\underset{\|}{C}}-Cl + NH_3 \longrightarrow$

(b) 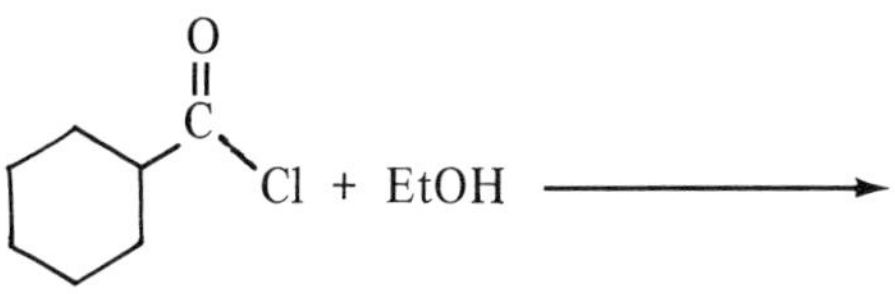 $\text{C}_6\text{H}_{11}-\overset{O}{\underset{\|}{C}}-Cl + EtOH \longrightarrow$

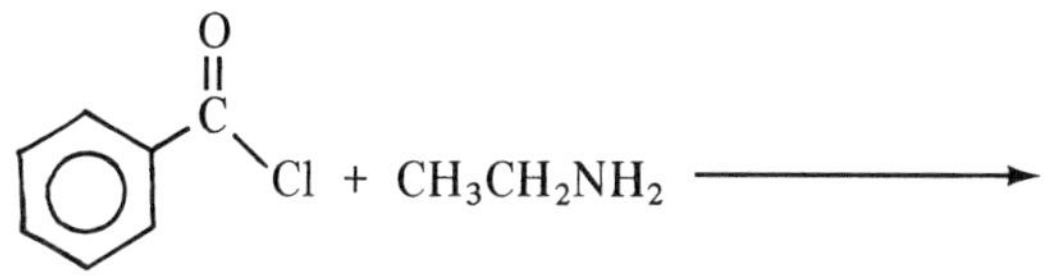

(c) $\text{Ph}-\overset{O}{\underset{\|}{C}}-Cl + CH_3CH_2NH_2 \longrightarrow$

(d) $CH_3-\overset{O}{\underset{\|}{C}}-Cl + CH_3-\overset{O}{\underset{\|}{C}}-O^{\ominus}Na^{\oplus} \longrightarrow$

Chapter 18
Acid Chlorides

*136. What are the products of each of the following reactions?

(a) $CH_3-\overset{O}{\underset{\|}{C}}-O-\overset{O}{\underset{\|}{C}}-CH_3 + H_2O \longrightarrow$

(b) $CH_3-\overset{O}{\underset{\|}{C}}-Cl + H_2O \longrightarrow$

(c) $CH_3-\overset{O}{\underset{\|}{C}}-NH_2 + H_2O \xrightarrow[\Delta]{H^{\oplus}}$

(d) $CH_3-\overset{O}{\underset{\|}{C}}-O-CH_3 + H_2O \xrightarrow[\Delta]{H^{\oplus}}$

Chapter 18
Carboxylic Acids

52 PROBLEMS

*137. Assuming you have α-naphthol, phosgene, and methylamine available, suggest a method of preparation for Carbaryl (1-naphthalenol methylcarbamate) without using methyl isocyanate. (See problem 160.)

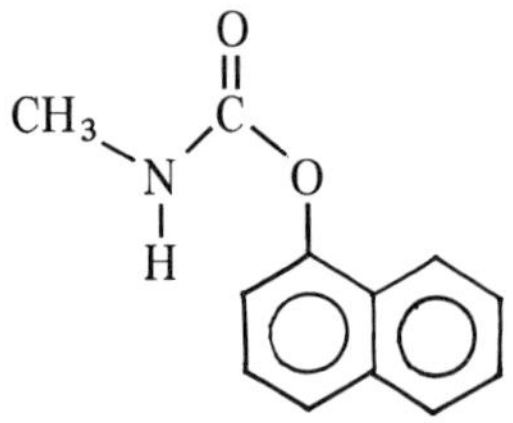

**Chapter 18
Reactions of Acid Chlorides**

*138. Gamma or δ-hydroxy carboxylic acids lose water to form lactones, but β-hydroxy carboxylic acids dehydrate to produce α,β-unsaturated acids. Show a mechanism for the formation of the lactone formed from γ-hydroxypentanoic acid.

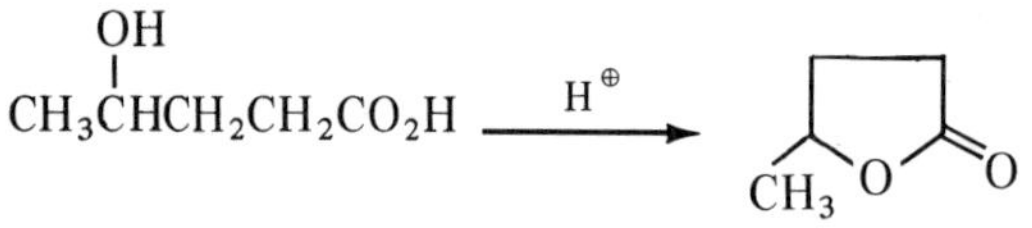

$$CH_3CHCH_2CH_2CO_2H \xrightarrow{H^\oplus}$$ (with OH on the first CH)

Why doesn't β-hydroxybutyric acid form a lactone?

**Chapter 18
Acids**

*139. What is the product? Suggest a possible mechanism.

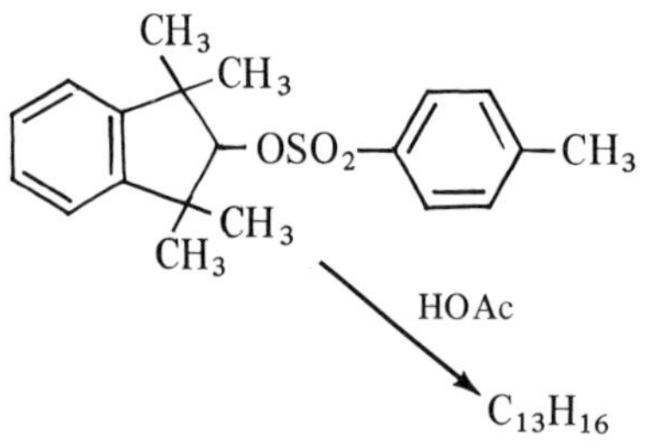

$C_{13}H_{16}$
NMR: δ 7.09 ppm (4H)
         1.98   (3H)
         1.82   (3H)
         1.15   (6H)
      (all are singlets)

**Chapters 15 and 18
Sulfonate Esters**

*140. Suggest a mechanism for the following:

$$\phi-CH_2CO_2Ag + Br_2 \xrightarrow[\Delta]{CCl_4} \phi-CH_2-Br + CO_2 + AgBr$$

**Chapter 18
Carboxylic Acids**

*141. In the following reaction, where is the labeled oxygen found in the product? Suggest a mechanism consistent with your answer.

$$\phi-\overset{O}{\underset{\|}{C}}-OH + CH_3CH_2-\overset{18}{O}H \xrightarrow{H^{\oplus}}$$

**Chapters 15 and 18
Acids and Alcohols**

*142. Upon opening a bottle of aspirin that was several months old and loosely capped, a student detected the smell of acetic acid. Account for these observations with verbal and mechanistic descriptions.

**Chapter 18
Esters**

*143. The conversion of a nitrile to a carboxylic acid is readily accomplished by acid catalyzed hydrolysis. Provide a mechanism for the conversion of benzonitrile to benzoic acid.

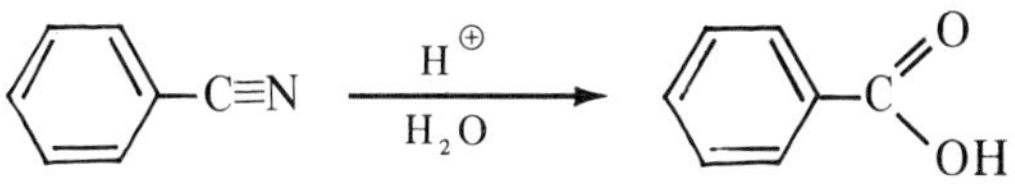

$$\phi-C\equiv N \xrightarrow[H_2O]{H^{\oplus}} \phi-C\overset{O}{\underset{OH}{\diagdown}}$$

**Chapter 18
Acids**

*144. Arrange the following in order of decreasing acidity. Justify your answer.

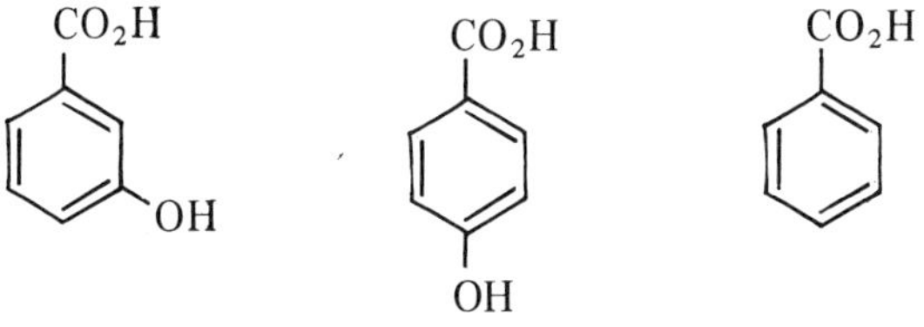

A: 3-hydroxybenzoic acid (CO₂H, meta-OH)
C: 4-hydroxybenzoic acid (CO₂H, para-OH)
(middle): benzoic acid (CO₂H)

**Chapter 18
Acids**

54 PROBLEMS

*145. How would you separate a mixture of the following? Assume they are dissolved in diethylether. Use formulas in your description.

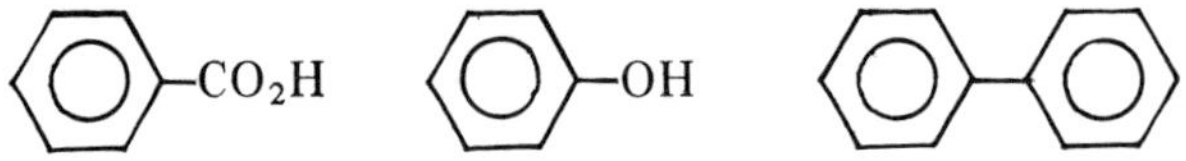

Ph—CO_2H Ph—OH   Ph—Ph

Chapters 15 and 18
Alcohols and Acids

**146. Explain why the pK_{a_1} for phthalic acid is smaller than the pK_{a_1} for terephthalic acid. Which acid would lose the *second* proton more easily? Why?

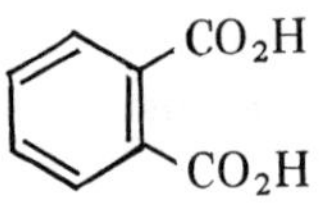

Phthalic acid (1,2-di-CO_2H benzene)   Terephthalic acid (1,4-di-CO_2H benzene)

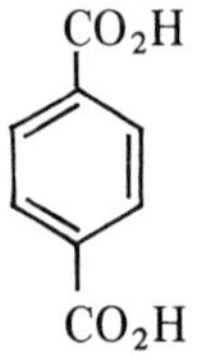

Chapter 18
Acids

**147. Suggest a mechanism for the following Beckmann rearrangement:

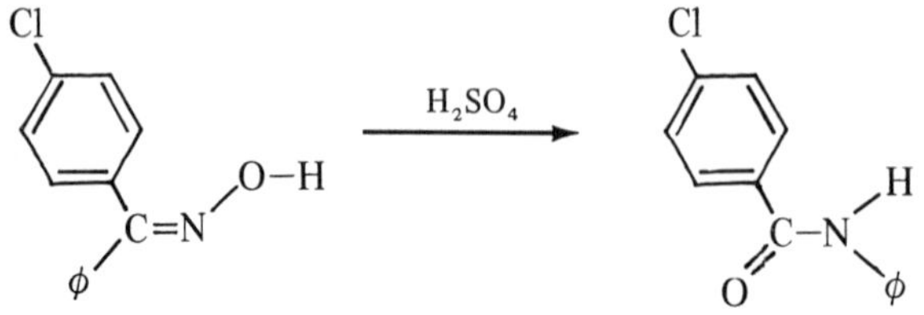

(4-Cl-C$_6$H$_4$)—C(φ)=N—O—H $\xrightarrow{H_2SO_4}$ (4-Cl-C$_6$H$_4$)—C(=O)—N(H)(φ)

Chapters 16 and 18
Acid Derivatives

*148. Arrange the following in order of acidity (greatest to least).

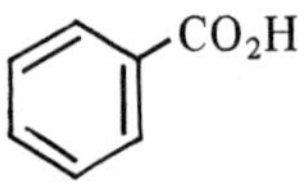

A: Ph—CO_2H B: 4-NO_2-C$_6$H$_4$—CO_2H C: 4-CH_3-C$_6$H$_4$—CO_2H

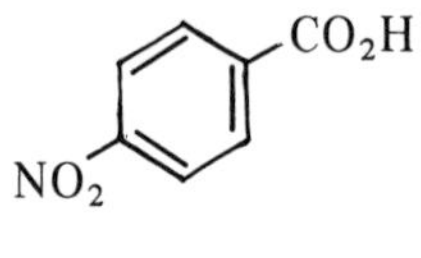

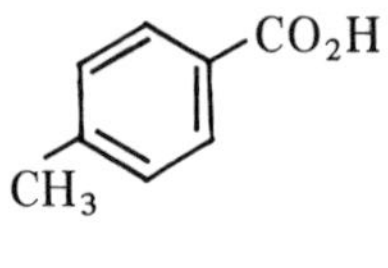

Chapter 18
Acids

PROBLEMS

**149. What is the product? What is the mechanism of the second step?

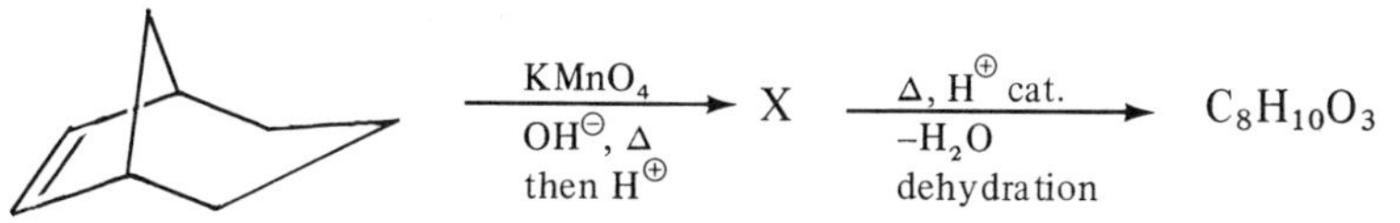

norbornene → (KMnO₄, OH⁻, Δ, then H⁺) → X → (Δ, H⁺ cat., −H₂O, dehydration) → C₈H₁₀O₃

Chapters 7 and 18
Alkenes and Acids

**150. Show a mechanism for the following reaction:

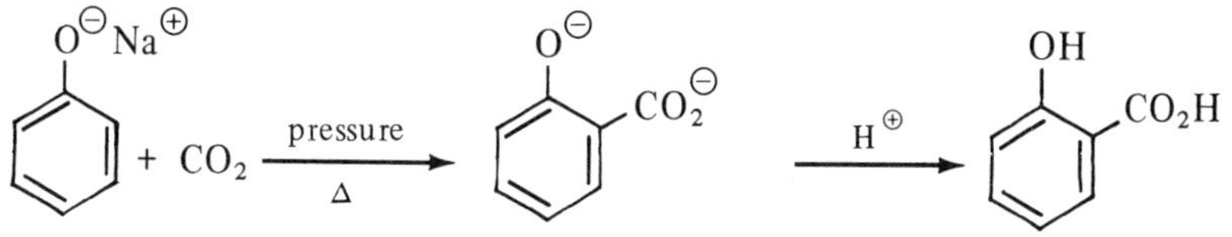

PhO⁻Na⁺ + CO₂ → (pressure, Δ) → 2-(O⁻)C₆H₄CO₂⁻ → (H⁺) → 2-hydroxybenzoic acid (salicylic acid)

Chapters 15 and 18
Phenols and Acids

**151. Diazomethane, CH_2N_2, is useful in the preparation of methyl esters of carboxylic acids. Draw resonance structures for diazomethane and suggest a mechanism that accounts for this reactivity.

$$R-\overset{O}{\underset{\|}{C}}-O-H + CH_2N_2 \longrightarrow R-\overset{O}{\underset{\|}{C}}-O-CH_3 + N_2$$

Chapter 18
Acids

**152. Starting with mesitylene, 1,3,5-trimethylbenzene, how would you synthesize 2,4,6-trimethylbenzoic acid?

(2,4,6-trimethylbenzoic acid structure)

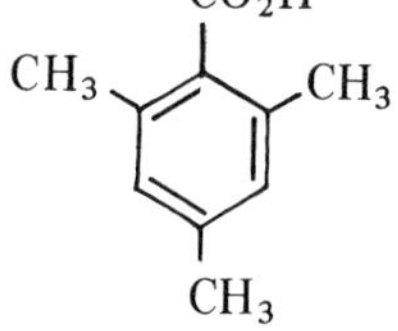

Chapter 18
Acids

153. Phthalic anhydride and benzene react in the presence of $AlCl_3$ to produce compound A, $C_{14}H_{10}O_3$. Compound A reacts with heat and sulfuric acid to produce compound B,

$C_{14}H_8O_2$. Compound B reacts with triphenylphosphine, bromomethane, and phenyllithium in THF to yield compound C, $C_{16}H_{12}$. Compound C reacts with diiodomethane and Zn(Cu) to produce compound D, $C_{18}H_{16}$. Propose structures for compounds A–D.

NMR information

Compound B	no absorption below 6 ppm
Compound C	two peaks with a 2:1 integration ratio
Compound D	two peaks with a 1:1 integration ratio

Chapters 16 and 18, Special Topic C
Acids

***154. Show a six-step synthetic sequence starting with adamantyl benzene (shown in the following reaction) that leads to the indicated product.

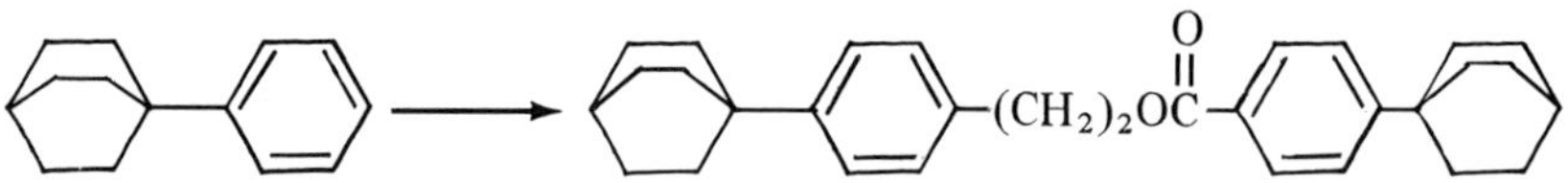

Chapters 12, 14, and 18
Acids

***155. A mixture of unknown organic materials was found in an unlabeled bottle in the refrigerator of an organic research lab. Chromatographic analysis showed the mixture to be primarily two compounds: A, an acidic material and B, a relatively neutral compound. The molecular weights of A and B are 178 and 162, respectively. Compound B on standing in air is slowly converted into compound A. Compound B gives a positive Tollens' test and readily forms a DNP derivative, whereas A does not. Treatment of compound A with thionyl chloride yields compound C, $C_{11}H_{13}OCl$, which when refluxed under Friedel–Crafts conditions (treatment with aluminum chloride in benzene) yields two α-tetralones, one with a methyl group adjacent to the ring fusion and peri to the carbonyl group, and another with the methyl group para to the carbonyl carbon atom. What are compounds A, B, and C and what is the sequence of reactions occurring?

Chapters 12 and 16–18
Acids

**156. Propose a mechanism for the following reaction:

[Structure: ethyl 1-(ethoxycarbonylmethyl)piperidine-4-carboxylate] →(1. NaOEt; 2. HCl)→ [bicyclic ammonium ketone with Cl⁻]

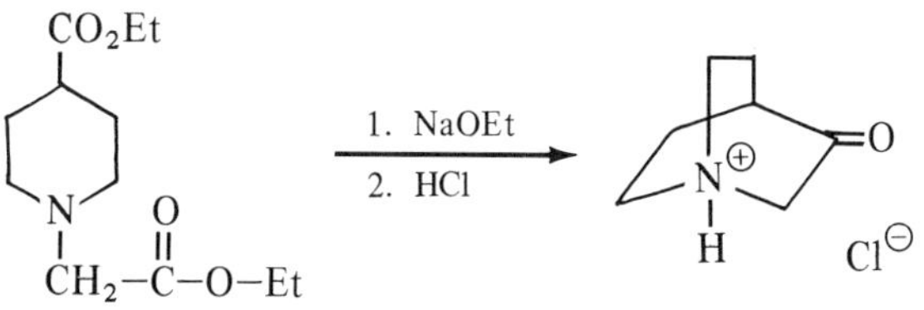

Chapters 18 and 19
Acids

*157. Write a mechanism for the following:

[1-(aminomethyl)cyclohexan-1-ol] →(NaNO$_2$ / HOAc)→ [cycloheptanone]

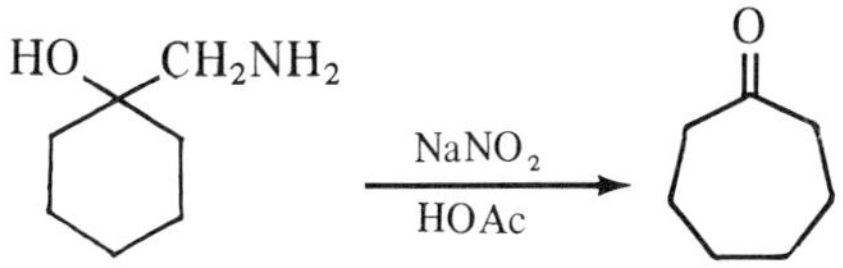

Chapters 15, 16, and 19
Amines

*158. The product of the following reaction is highly reactive. What is it? Why is it historically significant?

NH_2CH_3 + $COCl_2$ ⟶ ?

Methylamine   Phosgene   1 NMR peak

Chapter 19
Amines

*159. Show a synthetic route from aniline to benzoic acid that employs the Sandmeyer reaction as one step.

[PhNH$_2$] ⟶ ⟶ ⟶ [PhCO$_2$H]

Chapters 18 and 19
Amines and Acids

58 PROBLEMS

160. Carbaryl (an extremely effective insecticide) is formed as follows:

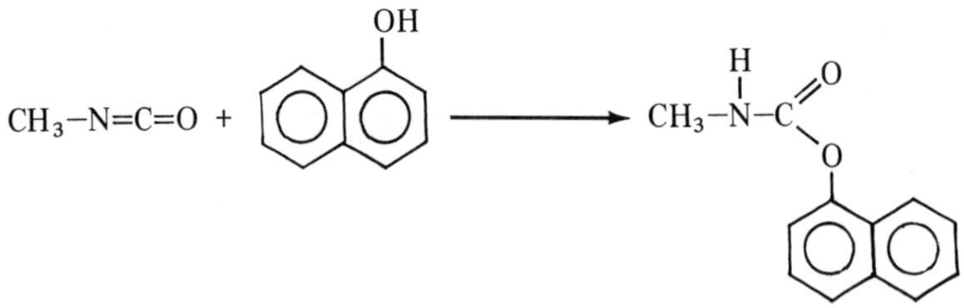

$CH_3-N=C=O$ + (1-naphthol) ⟶ Carbaryl (methylcarbamate ester of 1-naphthol)

Show the mechanism of Carbaryl formation.

Chapter 19
Isocyanates

*161. A synthesis of sulfathiazole, a sulfa drug, is outlined in the following sequence. What are compounds A–D?

$$C_6H_5NH_2 \xrightarrow{Ac_2O} \underset{A}{C_8H_9NO} \xrightarrow[-80°C]{HOSO_2Cl} \underset{B}{C_8H_8NO_3SCl}$$

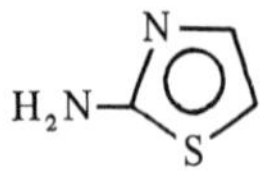

$$\xrightarrow{H_2N\text{-thiazole}} \underset{C}{C_{11}H_{11}N_3O_3S_2} \xrightarrow[\Delta]{dil.\ HCl} \underset{D}{C_9H_9N_3O_2S_2}$$

Chapter 19
Amines

*162. Suggest chemical tests that would distinguish between the following:

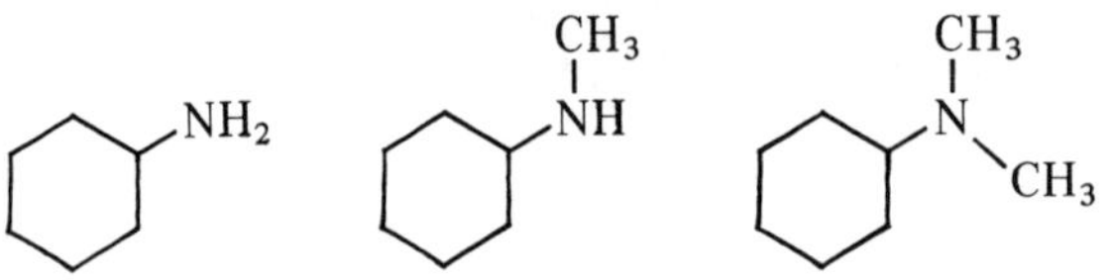

cyclohexyl-NH_2,   cyclohexyl-NH-CH_3,   cyclohexyl-$N(CH_3)_2$

Chapter 19
Amines

*163. Pyrrole readily reacts with electrophiles at the beta position, whereas pyridine does not. Explain.

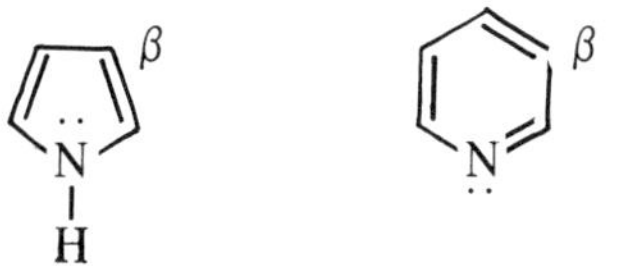

Chapter 19 and
Special Topic K
Amines

**164. What are compounds A, B, C, and D in the following sequence?

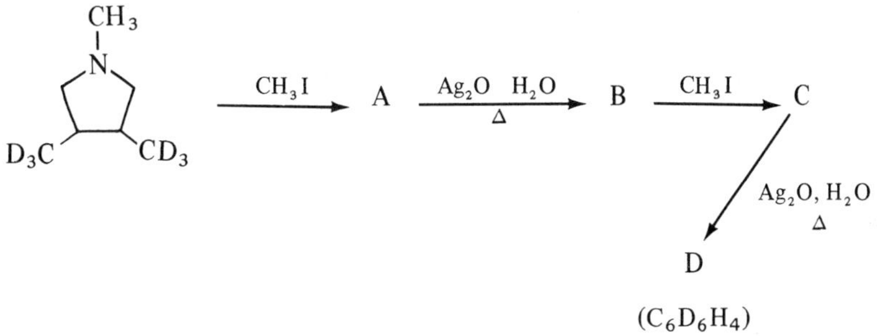

3,4-bis(trideuteromethyl)-1-methylpyrrolidine $\xrightarrow{CH_3I}$ A $\xrightarrow[\Delta]{Ag_2O\ H_2O}$ B $\xrightarrow{CH_3I}$ C $\xrightarrow[\Delta]{Ag_2O,\ H_2O}$ D ($C_6D_6H_4$)

Chapter 19
Amines

**165. Show how the following products are formed:

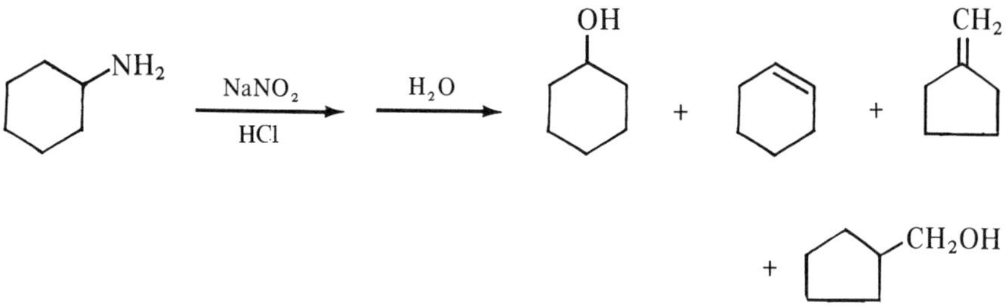

cyclohexylamine $\xrightarrow[HCl]{NaNO_2}$ $\xrightarrow{H_2O}$ cyclohexanol + cyclohexene + methylenecyclopentane + cyclopentylcarbinol

Chapter 19
Amines

166. Show a structure for the product and suggest a mechanism:

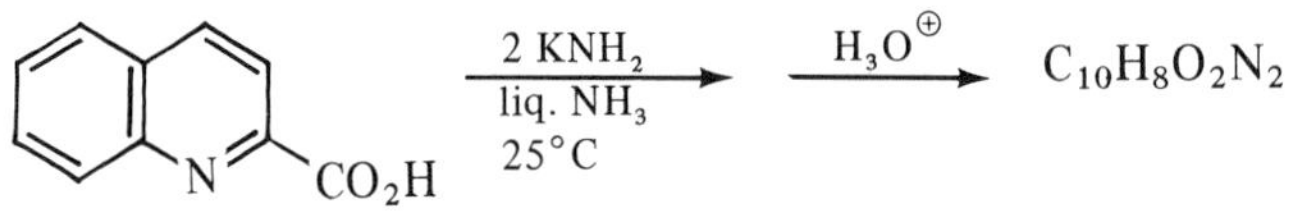

quinoline-2-carboxylic acid $\xrightarrow[\text{25°C}]{\underset{\text{liq. NH}_3}{2 \text{ KNH}_2}} \xrightarrow{H_3O^{\oplus}} C_{10}H_8O_2N_2$

Chapter 19
Amines

167. Compound A, $C_{13}H_{10}O$, reacts with hydroxylamine hydrochloride to produce compound B, $C_{13}H_{11}NO$, which melts at 141 °C. Compound A has one peak in the NMR spectrum and strong absorption just below 1700 cm^{-1} in the infrared spectrum. Compound B is soluble in aqueous base and gives a color with ferric chloride. When compound B is heated with acids (or PCl_5) it is transformed into compound C, $C_{13}H_{11}NO$, which melts at 163°C. Compound C is insoluble in aqueous base, but extended heating with NaOH produces benzoic acid and compound D, C_6H_7N, after work-up. Compound D reacts with acetyl chloride to produce compound E, C_8H_9NO, which melts between 112–114 °C. Propose structures for Compounds A–E.

Chapters 16, 18, and 19
Amines

168. A cyclic anhydride, compound A, reacts with ammonia to produce compound B, which has amide and ammonium carboxylate functionality. Heating compound B produces a new Compound, C, $C_8H_5NO_2$, plus water. Compound C reacts with alcoholic KOH to produce compound D, $C_8H_4NO_2K$. Compound D reacts with warm butyl bromide to produce compound E, $C_{12}H_{13}NO_2$, which upon heating with aqueous potassium hydroxide yields butylamine and another salt. Acid work-up of the salt yields compound F (neutralization equivalent = 83).

Basic hydrolysis of either compound A or E produces compound F. Show structures of Compounds A–F.

Chapter 19
Amines

169. Show a mechanism for the following:

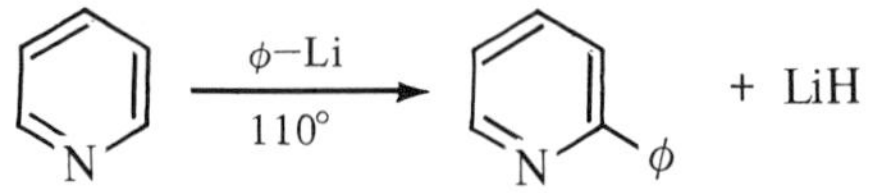

pyridine $\xrightarrow[110°]{\phi-\text{Li}}$ 2-phenylpyridine + LiH

Chapter 19
Amines

***170.** Write a mechanism that accounts for the following degradation:

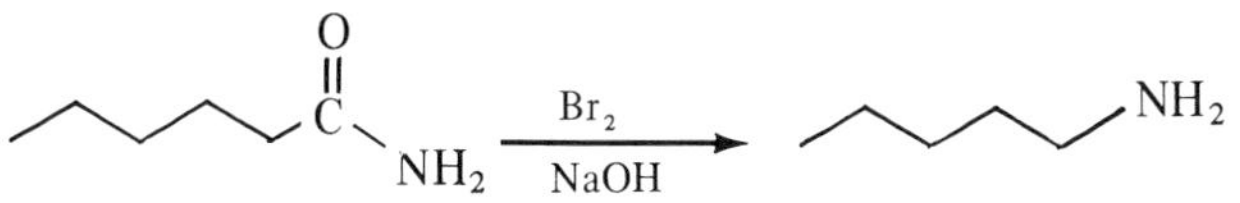

CH₃CH₂CH₂CH₂CH₂C(=O)NH₂ →[Br₂ / NaOH]→ CH₃CH₂CH₂CH₂CH₂NH₂

Chapter 19
Amines

***171.** Benzoic acid reacts with dicyclohexylcarbodiimide to yield an intermediate that reacts with propylamine to produce C₁₀H₁₃NO. Propose a structure for the product and suggest a mechanism for its formation.

Chapter 19
Amines

***172.** The Pictet–Spengler synthesis provides one route to 1-benzyltetrahydroisoquinolines. Suggest a mechanism.

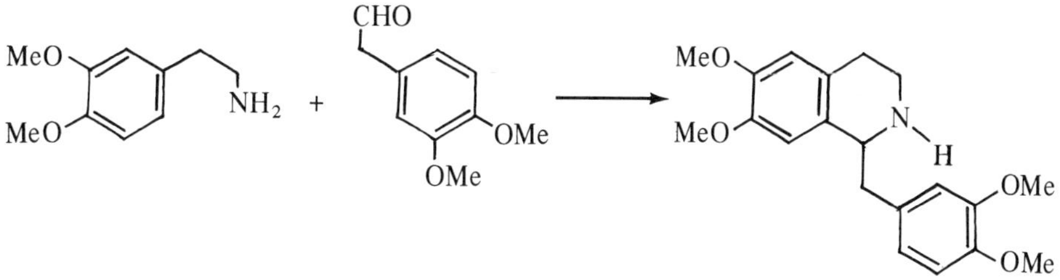

3,4-dimethoxyphenethylamine + 3,4-dimethoxyphenylacetaldehyde → 6,7-dimethoxy-1-(3,4-dimethoxybenzyl)-1,2,3,4-tetrahydroisoquinoline

Chapters 16 and 19
Amines

*173.** In a typical malonic ester or acetoacetic ester synthesis there is a decarboxylation in the final step. Write a mechanism for the decarboxylation of the following malonic acid derivative:

HO₂C–CH(CH₂CH₃)–CO₂H →[Δ]→ CH₃CH₂CH₂CO₂H + CO₂

Chapter 20
Esters

62 PROBLEMS

***174.** Suggest a mechanism for the following reaction:

4-bromobenzaldehyde + diethyl malonate $\xrightarrow[\text{EtOH}]{\text{Et}_2\text{NH}}$ 4-Br-C$_6$H$_4$-CH=C(CO$_2$Et)$_2$

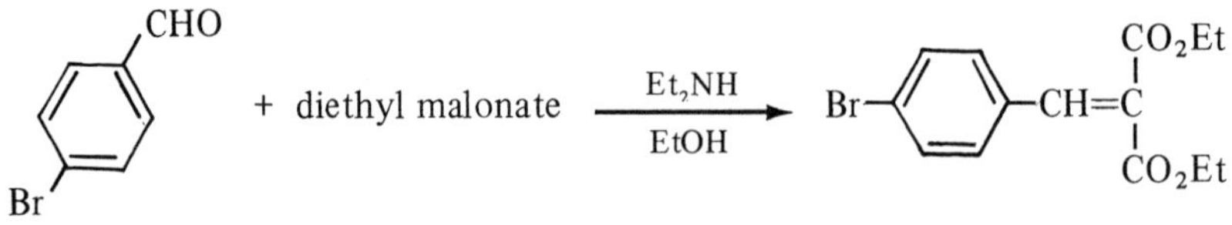

Chapter 20
Esters

**175. What is the most likely mechanism for the following transformation?

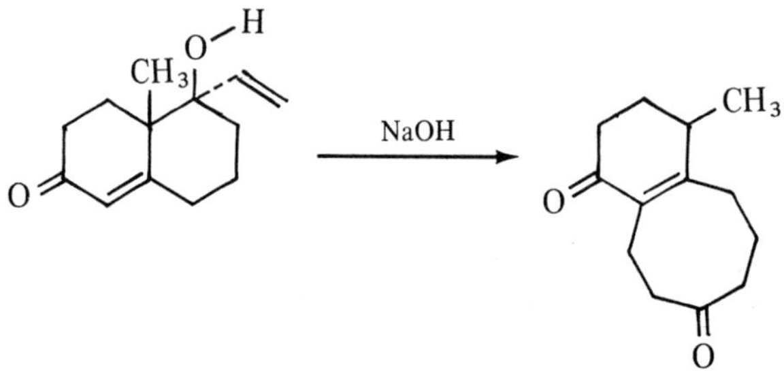

[bicyclic enone with CH$_3$, OH, and vinyl substituents] $\xrightarrow{\text{NaOH}}$ [ring-expanded bicyclic diketone with CH$_3$]

Chapters 17 and 20
Carbonyl Compounds

*176. Predict the product of the following reaction by supplying a reasonable mechanism.

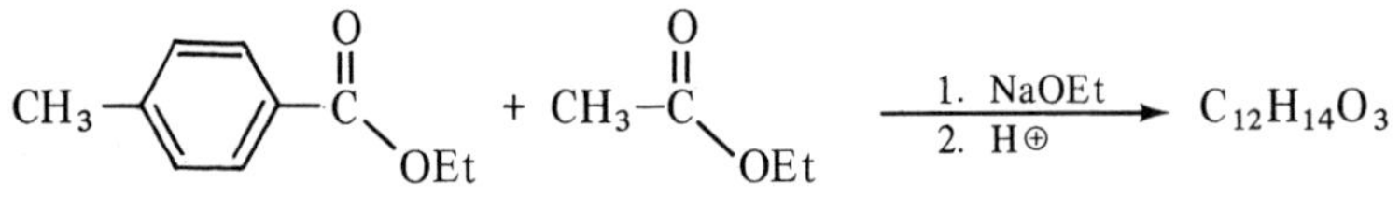

$$\text{CH}_3\text{-C}_6\text{H}_4\text{-C(=O)-OEt} + \text{CH}_3\text{-C(=O)-OEt} \xrightarrow[\text{2. H}^\oplus]{\text{1. NaOEt}} \text{C}_{12}\text{H}_{14}\text{O}_3$$

Chapter 20
Esters

**177. The sodium salt of diethylmalonate reacts with 1,3-dibromopropane to produce compound A, $C_{10}H_{17}O_4Br$. Compound A reacts with sodium ethoxide to produce compound B, $C_{10}H_{16}O_4$. Compound B produces compound C, a diacid, after acidification. Heating Compound C results in compound D, $C_5H_8O_2$. Suggest structures for compounds A, B, C, and D.

Chapter 20
Esters

178. Ethyl acetoacetate (acetoacetic ester) reacts with sodium ethoxide and bromoacetone to produce compound A, $C_9H_{14}O_3$. Compound A when refluxed in an aqueous solution of NaOH is converted to compound B, $C_7H_{10}O_4$, after an acid work-up. Upon heating, B is converted to compound C, $C_6H_{10}O_2$. The NMR spectrum of compound C shows no absorption above 6 ppm. What are the structures of compounds A, B, and C?

**Chapter 20
Esters**

179. Which of the following can be classified as reducing sugars?

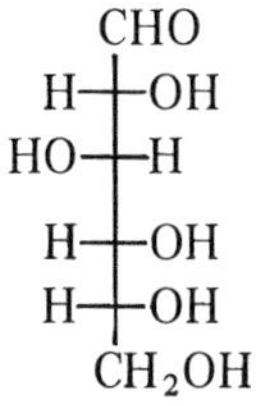

CHO
H—OH
HO—H
H—OH
H—OH
CH$_2$OH
A

B (pyranose with HOCH$_2$, OH groups)

C (pyranose with OCH$_3$ at anomeric position)

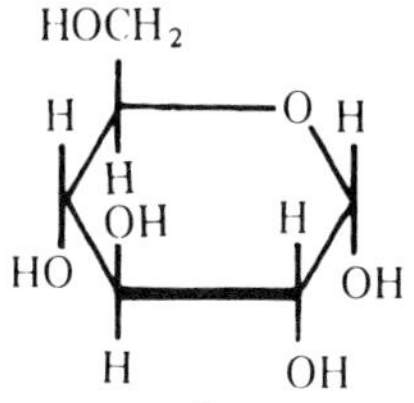

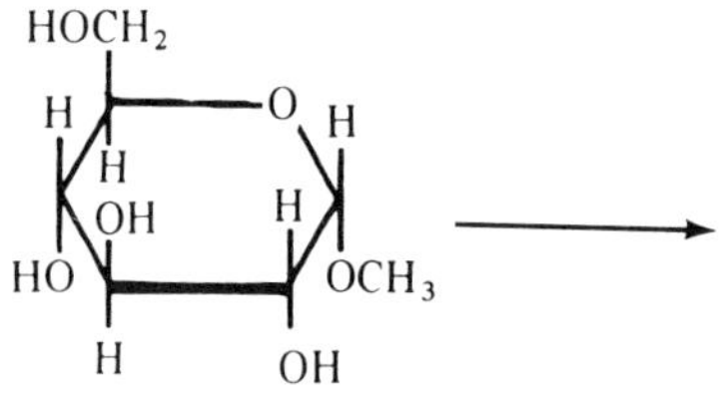

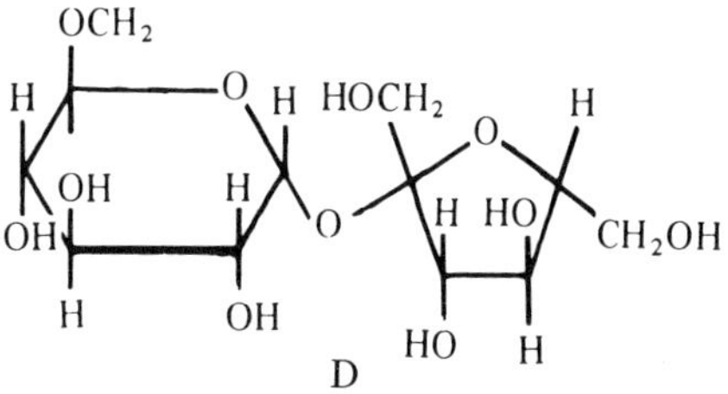

D (disaccharide structure with OCH$_2$, OH, HOCH$_2$, CH$_2$OH groups)

**Chapter 21
Carbohydrates**

180. One mole of sedoheptulose, $C_7H_{14}O_7$, undergoes oxidative cleavage when treated with aqueous periodic acid, HIO_4, to produce 1 mole of CO_2, 2 moles of formaldehyde, and 4 moles of formic acid. Selective degradation of this sugar produces the hexose D-altrose.

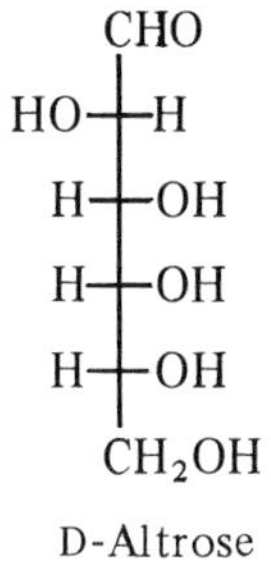

CHO
HO—H
H—OH
H—OH
H—OH
CH$_2$OH

D-Altrose

What is the structure of sedoheptulose? Justify your answer based on the preceding set of facts.

**Chapter 21
Carbohydrates**

64 PROBLEMS

*181. When one attempts to prepare a batch of fudge it is not uncommon to find a recipe that calls for a teaspoon of vinegar to be added to the sucrose solution just before the mixture is brought to a boil. Based on the following sweetness factors how does this make chemical sense? What reaction is taking place?

Sugar	Sweetness Factor
Glucose	1.00
Sucrose	1.45
Fructose	1.65

Chapter 21
Carbohydrates

***182. Aldoses and ketoses are unstable in basic solution where they isomerize. Propose a mechanism for the following interconversion.

D-Glucose $\xrightarrow{\text{NaOH}}$ D-Glyceraldehyde (A) + Dihydroxyacetone (B) $\rightleftharpoons$ D-Fructose

$$\begin{array}{c} \text{CHO} \\ \text{H}-\text{OH} \\ \text{HO}-\text{H} \\ \text{H}-\text{OH} \\ \text{H}-\text{OH} \\ \text{CH}_2\text{OH} \end{array} \xrightleftharpoons{\text{NaOH}} \begin{array}{c} \text{CHO} \\ \text{H}-\text{OH} \\ \text{CH}_2\text{OH} \end{array} + \begin{array}{c} \text{CH}_2\text{OH} \\ \text{C}=\text{O} \\ \text{CH}_2\text{OH} \end{array} \rightleftharpoons \begin{array}{c} \text{CH}_2\text{OH} \\ \text{C}=\text{O} \\ \text{HO}-\text{H} \\ \text{H}-\text{OH} \\ \text{H}-\text{OH} \\ \text{CH}_2\text{OH} \end{array}$$

D-Glucose   D-Glyceraldehyde (A)   Dihydroxyacetone (B)   D-Fructose

Chapter 21
Carbohydrates

**183. Which two diastereomeric aldohexoses give D-glucaric acid upon oxidation?

Chapter 21
Carbohydrates

*184. Complete hydrolysis of an unknown triglyceride produces the following acids and glycerol following work-up.

$CH_3(CH_2)_{12}CO_2H$

$CH_3(CH_2)_{14}CO_2H$

$CH_3(CH_2)_{16}CO_2H$

What are the possible structures for the triglyceride?

Chapter 22
Lipids

185. During electrophoresis at pH 1.5, which peptide migrates toward the cathode most rapidly? Why?

   I Ala-Glu-Val-Asp-Gly-Leu-Ala

   II Val-Lys-Ala-Leu-Phe-Lys-Leu

   III Gly-Leu-Val-Phe-Gly-Val-Val

Chapter 23 Proteins

186. In an attempt to determine the *N*-terminal amino acid of the following compound, preparation of a 2,4-dinitrofluorobenzene (DNFB) derivative failed to clearly establish the identity of the terminal residue.

Val-Ser-Ala-Leu-Phe-Orn-Leu

Why?

Chapter 23 Proteins

187. Cyanogen bromide, BrCN, is used to confirm the presence of methionine residues in proteins. Write a mechanism that illustrates the reaction of cyanogen bromide with methionine in the following peptide:

$$\begin{array}{c} CH_3 \\ | \\ S \\ | \\ CH_3 \quad\quad NH_2\ O \quad\ (CH_2)_2\ O \\ \diagdown \quad\quad\ |\quad\ || \quad\quad\ |\quad\quad || \\ CH-CH_2-CH-C-NH-CH-\!-C-NH-CH_2-CO_2H \\ \diagup \\ CH_3 \end{array}$$

Chapter 23 Proteins

188. In 1950 Pehr Edman published a method of *N*-terminal amino acid analysis of proteins that became widely used. Phenyl isothiocyanate is used to form a readily identifiable phenylthiohydantoin derivative of the *N*-terminal amino acid in the protein. Mild acid hydrolysis releases the rest of the protein unaltered so a *new N*-terminal group of the remaining peptide (one amino acid shorter) can be identified subsequently. Suggest a mechanism for the action of the Edman reagent (phenylisothiocyanate).

$$\phi-N=C=S \ +\ H_2N-\underset{\underset{R}{|}}{CH}-\overset{\overset{O}{||}}{C}-NH-\underset{\underset{R'}{|}}{CH}-\overset{\overset{O}{||}}{C}\!\sim\!\!\sim\text{etc.} \xrightarrow[\text{pH 9}]{^\ominus OH} \xrightarrow{H^\oplus}$$

$$\longrightarrow \text{phenylthiohydantoin}$$

Chapter 23 Proteins

*189. A synthesis of glycylalanine via the Bergmann synthesis is outlined in the following equation. Provide structures for compounds A–E.

$$H_2NCH_2CO_2H + \phi-CH_2OCOCl \longrightarrow C_{10}H_{11}NO_4 \xrightarrow{SOCl_2}$$
$$A$$

$$C_{10}H_{10}NO_3Cl \xrightarrow{alanine} C_{13}H_{16}N_2O_5Cl \xrightarrow{H_2/Pd} Gly\text{-}Ala + C_7H_8 + CO_2$$
$$B \qquad\qquad\qquad C \qquad\qquad\qquad D$$

**Chapters 23 and 18
Proteins**

***190. Amines, ammonia, and amino acids react with ninhydrin (triketohydrindene hydrate) to produce a purple compound. Show how this occurs for glycine.

**Chapters 19 and 23
Amino Acids and Proteins**

*191. Outline a mechanistic scheme for the phosphorylation of ribose.

**Chapters 15, 18, and 24
Nucleic Acids**

***192. Basic hydrolysis of RNA produces a mixture of 2' and 3' nucleotides, but DNA is unreactive under similar conditions. Show a mechanism for this hydrolysis.

**Chapter 24
Nucleic Acids**

68 PROBLEMS

*193. Starting with the *cis,trans*-2,4-hexadiene shown in the following reaction, arrange the four molecular orbitals of the pi system in order of increasing energy. From the signs of the orbitals of the HOMO (frontier orbital) in both the ground and excited states, predict the stereochemistry of the products for thermal and photochemically induced ring closure.

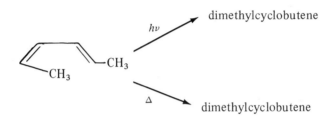

Chapter 24 and Special Topic N
Electrocyclic Reactions

**194. Consider the following thermally allowed intramolecular cycloaddition. Show how it occurs and where the deuterium in the product will be.

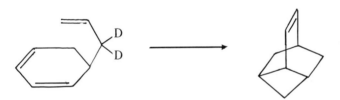

Chapter 24 and Special Topic N
Electrocyclic Reactions

***195. Does the following (thermal) reaction occur in a conrotatory or disrotatory mode? Explain in terms of frontier orbital interactions.

Would a different mode occur in the anionic cyclization? Explain.

Chapter 24 and Special Topic N
Electrocyclic Reactions

196. Using HOMO–LUMO interactions, predict whether the following cycloaddition reaction is thermally or photochemically allowed.

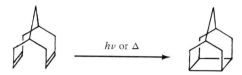

Chapter 24 and Special Topic N
Electrocyclic Reactions

197. What are Compounds A and B? Show mechanisms for these conversions.

$$\text{(structure)} \xrightarrow{300°C} C_7H_{12} \quad A$$
$$\xrightarrow{Br_2} C_7H_{12}Br_2 \quad B + C \text{ (isomers)}$$

This reaction may not be discussed in your book. Try it anyway.

198. What is the mechanism of the following reduction?

$$\text{2-methylcyclohex-2-enone} \xrightarrow[\text{2. } t\text{-BuOH work-up}]{\text{1. Li/NH}_3} \text{2-methylcyclohexanone}$$

This reaction may not be discussed in your book. Try it anyway.

199. A chemist attempted to brominate 2-ethylbenzoic acid with *N*-bromosuccinimide. Work-up of the reaction mixture on a very humid day and recrystallization from wet solvent produced a compound that gave a negative test for bromine in the sodium fusion test. The new compound was not acidic. The mass spectrum indicated a molecular weight of 148. The IR spectrum showed a carbonyl band consistent with the formation of an aromatic ester. The NMR spectrum showed four protons in the aromatic region. The upfield portion of the spectrum showed a sharp doublet that integrated for

70 PROBLEMS

three protons, and a quartet that integrated for one proton. What is the new compound? How did it form? Write mechanisms.

**Chapter 15
Alcohols**

***200.** An aromatic compound with the formula $C_9H_{10}O_2$ has the off-resonance proton decoupled ^{13}C NMR shown in the following spectrum. The proton spectrum of this compound shows an A_2B_2 pattern consistent with a para-disubstituted benzene. The compound gives a positive Tollens' test, and reacts with CH_3OH and H^+ to yield a new compound with formula $C_{10}H_{14}O_3$. This new compound shows no carbonyl absorption in the infrared. What is the initial compound? Assign all the peaks in the off-resonance decoupled spectrum.

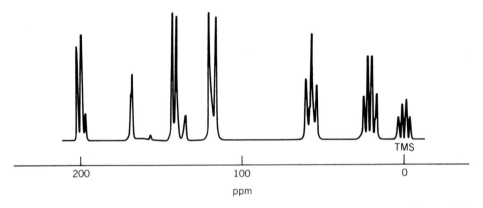

**Chapter 13
Spectroscopy**

***201.** The proton NMR spectrum of C_7H_8O shows three singlets at 7.35 (5H), 4.60 (2H), and 2.45 (1H). The off-resonance proton decoupled ^{13}C NMR spectrum shows absorptions at 64.5 (t), 127 (d), 128 (d), and 141 (s) ppm. What is the structure of this compound?

**Chapter 13
Spectrocopy**

***202.** The $^1J_{CH}$ coupling constant (i.e., between carbon and protons bonded directly to it) is proportional to the s character of the C–H bond in question. (i.e., $sp^3 < sp^2 < sp$). Based on this fact, match the following compounds with their respective C–H coupling constants, and then predict the percentage of s character in the bond between C and H in trifluoroacetaldehyde, which has a $^1J_{CH}$ of 207 Hz. Explain this value.

Match

CH$_3$—CH$_3$ 156 Hz
CH$_2$=CH$_2$ 248 Hz
CH≡CH   125 Hz

Chapter 13
Spectroscopy

***203. Compound A, C$_3$H$_4$O, has an off-resonance proton decoupled ^{13}C NMR spectrum showing a triplet at 50 ppm, a doublet at 74 ppm, and a singlet at 83 ppm. In the presence of a catalyst, compound A rapidly adds 2 mole of hydrogen to yield compound B, C$_3$H$_8$O. Careful reaction of compound B with acetyl chloride yields compound C, C$_5$H$_{10}$O$_2$, a pleasant smelling liquid. Pyrolyzing compound C at high temperatures over glass beads in the absence of oxygen yields propylene and acetic acid. What are the structures of compounds A, B, and C? Show the reactions described.

Chapter 13
Spectroscopy

***204. A compound, X, with molecular formula C$_{10}$H$_{20}$O, shows a very complex proton NMR spectrum, with all absorptions between about 1 and 4 ppm. The ^{13}C spectrum (off-resonance proton decoupled) shows 10 absorptions from 15 to 72 ppm. Three of these are quartets, four are doublets, and three are triplets. Under acidic conditions the compound loses H$_2$O and yields isomeric alkenes Y and Z (in unequal amounts). Hydrogenation of either Y or Z yields compound W, C$_{10}$H$_{20}$, a substituted cyclohexane. Compound W has a plane of symmetry. Reaction of X with sodium yields hydrogen and a sodium alkoxide. Compound X does *not* react with hydrogen at high pressure in the presence of a catalyst. Draw the structures of W, X, Y, and Z.

Chapter 13
Spectroscopy

**205. The ^{13}C relaxation time, T_1, nicely correlates with segmental freedom of motion, increasing as the degree of motion of molecular segments of a molecule increase. Consider the T_1 values for carbon atoms in decane and explain the trend in terms of molecular structure and your knowledge of how hydrocarbons change conformations.

CH$_3$ | CH$_2$ | CH$_2$ | CH$_2$ | CH$_2$ | CH$_2$ | CH$_2$ | CH$_2$ | CH$_2$ | CH$_3$

T_1:   8.7 | 6.6 | 5.7 | 5.0 | 4.4 | 4.4 | 5.0 | 5.7 | 6.6 | 8.7

Chapter 13
Spectroscopy

3 VERBAL LEADS

1. Shared electron pairs are represented with bars (—); unshared electron pairs are noted with dots. Recall that the formula for the calculation of formal charge on any individual atom involves the valence electrons.

   $FC = Z -$ (number of unshared individual electrons $+ \frac{1}{2}$ the number of shared electrons)

   Z refers to the group number in the Periodic Table. For example, $Z = 5$ for N, $Z = 6$ for O, and so on.

2. The percentages do not add up to 100. Assume the difference is due to the presence of oxygen in the compound:

   $100 - (40.01 + 6.11) = 53.28\%$ oxygen

   Divide the percentage composition of each element by each corresponding atomic mass. Find the smallest whole number ratio between the elements. This operation yields the *empirical* formula. To determine the molecular formula, divide the molecular mass by the mass of the empirical formula and then multiply the empirical formula by this factor.

3. Remember that free rotation exists around single (sigma) bonds. Normally carbon, oxygen, and hydrogen have four, two, and one bond, respectively. It should be noted that any two-dimensional structure here is only an approximation. Bond angles, corresponding to various hybridizations of the carbon atoms provide a more accurate description of the molecular geometry. The following two-dimensional structural representations are identical.

   CH_3CH and $CH_3CH_2CH_3$
   $\quad\ \ |$
   $\ \ \ CH_3$

4. Consider the hybridization of each carbon atom. Note that carbon with four single bonds is sp^3 hybridized, carbon with one double bond is sp^2 hybridized, and carbon with one triple bond is sp hybridized. The different geometric arrangements are:

sp	linear
*sp*²	planar
*sp*³	tetrahedral

5. Consider the hybridization of each of the central atoms involved. It might help to redraw the molecule with attention to molecular shape and orbital arrangement resulting from hybridization of the central atom. For example, sp = linear, sp^2 = planar, and sp^3 = tetrahedral. Unshared pairs of electrons make large contributions to the dipole moment.

6. Consider the different types of arrows. Can you identify the different species? In (a) note the position of the electrons; in (b) note the arrangement of the atoms. Resonance structures result from a change in placement of *electrons* in the Lewis structures. Isomers result from the repositioning of *atoms*.

7. Look for subtle differences in the structures. Consider the influence of the functional group on boiling point. Consider the potential for hydrogen bonding, differences in molecular weights, and differences in dipole moments. Recall that for hydrogen bonding to occur, hydrogen must be covalently bonded to N, O, or F. If functional groups are identical, then molecular weight differences become more important in determining boiling points. Generally, the greater the molecular weight, the greater the boiling point if the structures are similar. Other things being equal, the greater the dipole moment, the higher the boiling point.

8. Recall that polar solutes dissolve in polar solvents. Like dissolves like. Ether is of low polarity, but both water and sulfuric acid are highly polar. At first glance then, diethyl ether shouldn't dissolve in either solvent. Consider the differences in acidity between sulfuric acid and water. Can the stronger acid protonate the oxygen of the ether? The oxygen of the ether is protonated by sulfuric acid. The resulting ion pair is soluble in the unreacted sulfuric acid that acts as a solvent.

9. Recall that,

$$K_a = \frac{[H^\oplus][A^\ominus]}{[HA]}$$

where K_a is the dissociation constant for a weak acid. Weaker acids have smaller K_a's; stronger acids have larger K_a's. Note also that pK_a = -log K_a. In other words, the stronger the acid, the lower the pK_a. Consider the structural differences between the two acids. Which acid is more likely to lose a proton? Due to the electronegativity of the fluorine, trifluoracetic acid loses its protons more easily. In addition, the conjugate base of CF_3CO_2H, $CF_3CO_2^{\ominus}$, is better stabilized than the acetate ($CH_3CO_2^{\ominus}$) due to the inductive effect of the fluorine atoms.

10. Assume carbon has four bonds, oxygen two, and hydrogen one. The carbon chain *can* be branched. There are three isomers; one of these is an ether.

11. A typical ether has the general structure R—O—R. Consider both symmetrical ethers (R=R) and unsymmetrical ethers (R≠R). How many variations can you produce for R? Avoid duplications. There are 11 ethers.

12. Consider the use of a carbon to oxygen *double* bond in your answer. Draw an acceptable resonance structure for the acetate ion. Use the resonance concept to explain the equal carbon to oxygen bond lengths.

13. Consider branching possibilities. Note

$$CH_3CHCH_2CH_3 \quad \text{and} \quad CH_3CH_2CHCH_3$$
$$||$$
$$ClCl$$

are *not* different. Four isomers are possible for the formula C_4H_9Cl.

14. The ring exists in the chair conformation. The larger group (*t*-butyl) occupies the equatorial position. In a 1,3-disubstituted cyclohexane, one substituent is axial, the other equatorial. Since chlorine is *trans* to the *t*-butyl group, it must occupy the axial position.

15. Consider the Corey–House (Corey–Posner) synthesis. (a) Prepare bromocyclohexane by heating cyclohexane with bromine. (b) Next, prepare bromoethane by heating bromine with ethane. Bromocyclohexane and lithium produce cyclohexyllithium. Cyclohexyllithium can be reacted with cuprous iodide to produce lithium dicyclohexylcuprate. This cuprate can be treated with ethyl bromide to produce the product. An alternative route using cyclohexyl bromide *in the last step* is less attractive. Why?

16. As a base name, use the name of the alkane corresponding to the total number of carbon atoms *in the two fused* rings. For example, (a) = bicyclo[???]hexane. The carbon atoms common to both rings are named bridgeheads. Each chain of carbon atoms connecting the bridgeheads is called a bridge. (In specifying the length of the bridge, do *not* count the two bridgehead carbon atoms.) If a substituent is present on the bicyclic system, begin numbering at the bridgehead and proceed through the longest bridge first, then through the second longest bridge back to the first bridgehead.

17. Select the longest continuous chain to which the hydroxyl is directly attached. Drop the *e* on the name of the parent alkane and add *-ol*. Number the longest continuous carbon chain so that the carbon atom attached to the hydroxyl group has the lowest possible number. Indicate the position of other substituents by using numbers corresponding to the position of the substituent on the chain.

18. After choosing the longest chain, begin numbering at the end nearest a substituent. (The horizontal line of atoms does not always represent the longest chain.) Use numbers to identify the position of the substituents. When two or more of the same substituents are present, use the prefixes di-, tri-, tetra-, and so on. When two substituents are present on the same carbon atom, use the number twice. Groups are listed alphabetically.

19. Consider decalin to be an extension of a 1,2-disubstituted cyclohexane in the chair conformation. (Working with a model helps.) *trans*-1,2-Disubstitution requires either a diaxial or a diequatorial arrangement on the parent cyclohexane. The two hydrogen atoms at the ring fusion *must* both be axial; the four carbon chain is not long enough to connect the bridgeheads via the diaxial arrangement. Therefore, the second ring is formed by replacing two equatorial hydrogen atoms on the parent cyclohexane.

20. The reaction must occur in two stages. In the first stage chlorocyclopentane must form initially as follows:

$$C_5H_{10} + Cl_2 \longrightarrow C_5H_9Cl + HCl$$

Look at the structure of chlorocyclopentane and see how many different kinds of hydrogen atoms there are in this molecule. Each one of these different hydrogen atoms can be replaced by chlorine in the second chlorination, that is, of chlorocyclopentane. There are five different kinds of hydrogen atoms on chlorocyclopentane. The second chlorine can become bonded to the same carbon as the first chlorine, or to the adjacent carbon (C-2) in both a cis and trans fashion, or to C-3, in both a cis and trans fashion.

21. The compound cannot contain a double bond because permanganate is not decolorized, so it must contain a single ring. There are three possible single ring structures of formula C_5H_{10}, and only one of these can yield a single type of monochloro product. The possible compounds are *cis*-dimethylcyclopropane, *trans*-dimethylcyclopropane, methylcyclobutane, and cyclopentane. Only cyclopentane has a single type of replacable hydrogen.

22. The reaction must have a chain initiating step. In this instance it is the homolytic dissociation of chlorine, $Cl_2 \xrightarrow{h\nu} 2Cl\cdot$. This is followed by two chain propagating steps. The first involves hydrogen abstraction from ethane by the chlorine radical to produce an ethyl radical and one of the products, HCl. The second chain propagating step involves attack of the ethyl radical on Cl_2 to produce the product CH_3CH_2Cl and the chain propagating chlorine radical $Cl\cdot$. Chain termination steps can be a combination of any two radical intermediates.

23. $E_{act} = \Delta H^0$ when bonds are boken but none are formed, therefore the reaction $Br_2 \longrightarrow 2Br\cdot$ is characterized by the right-hand reaction diagram. The left-hand diagram thus represents the reaction of bromine radicals with methane. A, B, and C represent $CH_4 + Br\cdot$, $\overset{\delta\cdot}{Br}\cdots\overset{\delta\cdot}{H}-CH_3$, and $HBr + CH_3\cdot$, respectively, whereas D and E represent Br_2 and $2Br\cdot$, respectively. The vertical distance X–Y represents ΔH^0 for the abstraction reaction. The vertical distance X–Z represents E_{act} for the abstraction reaction. The vertical distance V–W represents ΔH^0 and E_{act} for the homolytic dissociation of Br_2.

24. Consider the mechanism of free radical bromination of an alkane, including initiation, propagation, and termination steps. Each product requires a different intermediate. Isopropyl bromide results from $CH_3\dot{C}HCH_3$ and propyl bromide from $CH_3CH_2CH_2\cdot$. Calculate ΔH for each of the two endothermic propagation steps by using bond dissociation energy tables. Compare these two values.

$Br\cdot + CH_3CH_2CH_2-H \longrightarrow CH_3CH_2CH_2\cdot + HBr \quad \Delta H = + 10.5 \text{ kcal/mol}$

(43.9 kJ/mol)

$Br\cdot + CH_3\underset{H}{\overset{|}{CH}}-CH_3 \longrightarrow CH_3\dot{C}HCH_3 + HBr \quad \Delta H = + 7 \text{ kcal/mol}$

(29.3 kJ/mol)

The higher the ΔH (more positive), the greater the energy needed to bring about a reaction. The thermodynamics of the rest of the chain reaction mechanism can be ignored because the other steps are either exothermic or identical in energy to the proton abstraction step.

25. Because the two methyl groups are equivalent, it is important to avoid double counting. There are four monochlorination products but only two monobromination products. Consider the selectivity of bromine versus chlorine in free radical halogenation.

26. Bromine is less reactive than fluorine or chlorine and is therefore a very selective reagent, reacting only with the most labile C–H bond. At the transition state for the abstraction process, radical character at the carbon center under attack is considerable

(*much bond breaking has occurred*). This radical character is best stabilized at the benzylic position. The product is 2-bromo-2-phenylpropane. The transition state for a reaction leading to 1-bromo-2-phenylpropane is much higher in energy.

27. A negatively charged nucleophile is always stronger than its conjugate acid. Therefore $^\ominus$OH is stronger than H_2O, and so on. Nucleophilicities parallel basicities as long as the nucleophilic atom in the group of nucleophiles is the same. The strongest base here is the methoxide ion; the weakest base is H_2O.

28. This reaction is an example of protolysis. It begins by donation of a proton from $FSO_3H \cdot SbF_5$ to the alkane. The intermediate species formed has a pentacoordinated carbon atom, involving a three-center bond. The intermediate loses H_2 giving the resultant product, diphenylethyl cation. The counter ion is $SbF_6^\ominus$.

29. Note that doubling the concentration of hydroxide ion does not affect the reaction rate. Doubling the concentration of RBr doubles the rate of the reaction. The reaction is therefore first order in RBr and zero order in $^\ominus$OH. The mechanism involves two steps. The first is a slow ionization of RBr to $R^\oplus$ and $Br^\ominus$. The second is a fast recombination of $R^\oplus$ and $^\ominus$OH to yield the product.

30. Formic acid is a very nonnucleophilic solvent. Bromine departs heterolytically (ionizes) from the substrate to yield a carbocation and bromide ion. The carbocation is attacked by formic acid to yield an adduct. Loss of a proton from this adduct to bromide ion yields the ester product and HBr. This is a typical S_N1 solvolytic reaction of a tertiary substrate. An elimination product would probably also form.

31. Nucleophilic attack by thiol sulfur occurs on the phenylethyl tosylate benzylic carbon. Inversion of configuration occurs at this carbon in the resultant protonated thiol substitution product. Loss of the proton bonded to positive sulfur in the initial substitution product to *p*-toluenesulfonate yields the methyl-1-phenylethyl sulfide product, along with *p*-toluenesulfonic acid.

32. The conditions ($I^\ominus$, acetone) are typical for S_N2 displacement. In the conformation of the cyclohexyl chloride shown, however, backside S_N2 attack is hindered by the axial methyl groups. Conformational flip to the other chair conformation puts the Cl atom axial and removes backside hindrance of the axial methyls. Backside S_N2 attack by iodide ion on the carbon atom bearing chlorine results in inversion at this center, eventually yielding the cyclohexyl iodide product of inverted configuration. The iodine atom is cis to the 1,3-dimethyl groups; chlorine in the starting material is trans.

33. In Reaction (1) the bromine in CH_3CH_2Br is not a particularly good leaving group and

attack by the oxygen of CH_3CH_2OH on the carbon atom bearing bromine is not very rapid. Silver ion can coordinate with $Br^{\ominus}$, accelerating its departure (increasing its stability as it departs by allowing silver bromide to form instead of $Br^{\ominus}$). A similar function is played by $H^{\oplus}$ in Reaction (2) where $-OH$ is converted to the better leaving group $-\overset{\oplus}{O}H_2$, which can depart as water ($^{\ominus}OH$ is a very poor leaving group). The products are, respectively, diethyl ether and ethyl bromide.

34. Sodium ethoxide is a very strong base that promotes elimination reactions. There are two *different* hydrogen atoms β to the bromine that are trans antiperiplanar. Abstraction of these two hydrogen atoms in a typical E2 elimination yields two products, 3-methylcyclohexene and 4-methylcyclohexene.

35. This is a solvolysis reaction of a tertiary substrate. The *t*-butyl chloride ionizes to give the *t*-butyl cation, which proceeds via several routes to the possible products. The cation can be captured by three nucleophiles present in the mixture; chloride ion, water, or methanol. Capture by chloride ion yields the starting material, whereas capture by water or methanol yields an alcohol or ether, respectively. Loss of a hydrogen on a carbon atom adjacent to the positive center (to the solvent: CH_3OH or H_2O) yields an elimination product.

36. This reaction yields an organic product with no bromine. The substrate is very hindered and not susceptable to backside S_N2 displacement. The nucleophile is strongly basic and quite bulky. It is most effective at proton abstraction and least effective as a nucleophile, especially with such a hindered substrate. Proton abstraction at a β carbon with concomitant departure of bromide ion yields the E2 elimination product.

37. This is a typical nucleophilic substitution reaction. The nucleophile is the amine and the electrophilic substrate undergoing nucleophilic attack is the sulfonium salt. S_N2 attack by the nitrogen of the amine on the aliphatic carbon atom adjacent to the positive charge in the sulfonium salt displaces diphenylsulfide to give the inverted product, a quaternary ammonium compound.

38. Both reactions proceed by similar mechanisms. In each case the leaving group departs in an S_N1 fashion in the slow step. Since the leaving group in Reaction (1) is *different* from that in Reaction (2), the ionization of each substrate will occur at a different rate. Different bonds are broken in the respective slow steps. The resultant carbocation is the same in both Reactions (1) and (2), and it will be captured in a fast step in Reaction (1) in exactly the same way as in Reaction (2). The factors (solvation, steric hindrance, etc.) that partition the product to alcohol or ether will be exactly the same in Reactions (1) and (2) because these reactions proceed through a common intermediate.

VERBAL LEADS 79

39. Formic acid (HCO$_2$H) is a very polar ionizing solvent. *p*-Nitrobenzoate is a good leaving group on a primary carbon atom that is subject to nucleophilic attack. The sulfhydryl group SH is strategically located with respect to the primary CH$_2$—OPNB function. Intramolecular nucleophilic attack by the SH function on the carbon atom bearing —OPNB occurs to give a cyclic sulfide.

40. Two of these ions are primary carbonium ions and one is a secondary carbonium ion. In simple alkyl systems carbonium ion stability in the order tertiary>secondary>primary might lead one to suppose that A would be the most stable. Other functionality is present here, however. The positive end of the C=O dipole is adjacent to the positive charge in C, so this species is not very stable, as adjacent like charges are not favored. The stabilizing inductive effect of the methyl groups in A is much less than that of the nitrogen lone pair, which can stabilize B.

41. The reactions involve elimination of HBr from the tertiary bromo substrate. Reaction (1) is a typical E2 (bimolecular) elimination involving a strong unhindered base, whereas reaction (2) involves a strong and more hindered base. Methoxide in reaction (1) removes the hydrogen leading to the most stable, highly substituted alkene. This reaction predominates because some of this alkene stability is available at the transition state for elimination, favoring this path. Bulky *t*-butoxide abstracts the least hindered hydrogen to yield the least substituted olefin. This path predominates for Reaction (2).

42. The hydroxyl function is protonated by sulfuric acid. Water then departs. Concurrent with departure of water is a shift of an adjacent methyl group to the primary position. The resultant tertiary carbonium ion loses a β proton to yield an olefin. The β proton can be lost from the adjacent methylene or from the adjacent methyl, to give the internal olefin (major product) or the terminal olefin (minor product), respectively.

43. The formula for a hydrocarbon with no ring or double bond is C_nH_{2n+2}. Since the original compound absorbed 1 mol of hydrogen, it must contain one double bond. The hydrogenation product, C_7H_{14}, must contain one ring since the total number of C—H bonds is two less than maximum. The difference between the maximum number of hydrogen atoms possible and the number of hydrogen atoms in this product is due to the presence of a carbocyclic ring.

44. Imagine the structure on the left resting on a table with the H, Cl, and Br touching the table top. Tip the structure over, lifting the H "off the table," while the Br and Cl remain "on the table" and the CH$_3$ goes over and down to touch the table on the other side of the Cl and Br. Now the Br, Cl, and CH$_3$ touch the table top (Cl out front,

Br in back with H and Cl in the plane). Keeping H where it is, rotate around the C—H axis to reach the structure on the right.

45. Reagent X is zinc and acetic acid. This reagent removes both bromines from the starting material to yield the olefin and $ZnBr_2$. Addition of HBr to the olefin yields an additional product, the tertiary bromide, formed via the most stable tertiary carbocation. Hydrolysis of this tertiary bromide in an aqueous solvent system will result in an S_N1 solvolysis, yielding the tertiary alcohol product.

46. Each reaction begins with electrophilic addition to a carbon—carbon double bond. The electrophilic species are $H^\oplus$, $H^\oplus$, Br·, $^\oplus O-O-O^\ominus$, and $Br^\oplus$, for reactions (a)—(e), respectively. Products in reactions (a), (b), and (e) result from combination of the carbocation intermediate with $Cl^\ominus$, $HSO_4^\ominus$, and H_2O, respectively. The anti-Markovnikov bromide (1-bromopentane) forms in reaction (c) via attack of the free radical intermediate on unreacted HBr. In reaction (d) cleavage of the ozonide produces formaldehyde and butanal.

47. Protonation of the starting alkene yields a tertiary carbocation. This reacts with another equivalent of alkene in a dimerization to yield a second tertiary carbocation. The 2-pentene is then formed by loss of a C-3 proton or C-3 deuterium from this second cation to give the product olefin. If $H^\oplus$ is lost the product will have a deuterium on C-3 and on the 4-methyl group. If $D^\oplus$ is lost then deuterium will appear only on the 4-methyl group. The 1-pentene is formed by loss of a proton from the C-1 methyl group. The 1-pentene product will have a deuterium at C-3 and C-5.

48. The double bond attacks the electrophilic mercuric acetate to yield a mercury substituted carbocation. This cation is attacked by methanol to yield a protonated (methoxycyclohexyl) mercuric acetate. Loss of the proton on the methoxyl oxygen yields X. Attack of hydride from borohydride on the carbon atom bearing mercury displaces elemental mercury and acetate to yield Y, methoxycyclohexane.

49. Diborane, $(BH_3)_2$, acting as if it were monomeric, that is, BH_3, adds across the double bond in a concerted fashion, boron going to the least hindered and hydrogen to the most hindered carbon atom. The two B—H bonds in this initial adduct add to two additional equivalents of $(CH_3)_2C=CH_2$ to give triisobutylborane, X. Triisobutylborane is hydrolyzed and oxidized by $H_2O_2/^\ominus OH$ to 2-methyl-1-propanol, Y.

50. Hydroboration of the double bond of Compound X yields tri-(cyclohexylmethyl) borane. Decomposition of this borane is hot deuterioacetic acid yields methylcyclohexane bearing the deuterium on the methyl group where boron was bonded in the borane. Reduction of Compound X by H_2 in the presence of a platinum catalyst yields methylcyclohexane. Using D_2 in this last reaction yields methylcyclohexane with a deuterium on the methyl group and on the tertiary ring carbon.

51. Two carbonyl groups are formed from the ozonization of a carbon–carbon double bond. Possible candidates for the alkenes can be formed "on paper" by changing two carbonyl groups (C=O + O=C) to C=C. Ozonization that results in only one product requires a symmetrical olefin (alkene).

52. The electron rich double bond of the olefin attacks the electrophilic bromine displacing bromide ion. A bromonium ion is formed. The bromonium ion "ring" is opened up via backside attack by bromide ion to give a vicinal dibromide. Bromide attack on the bromonium ion can occur at either the carbon atom bearing a methyl group or the carbon atom bearing an ethyl group. The result is the production of an enantiomorphic pair. Based on this description each enantiomer could be formed in different amounts because the transition states for opening the bromonium ion are not of equal energy. However, the bromonium ion has a mirror image that also forms, and when it opens up from bromide attack on the carbon atom bearing a methyl group, the product is identical to that formed by attacking its mirror image on a carbon atom bearing an ethyl group. The result is a racemic form.

53. The products result from addition to the double bond. Consider the mechanisms involved. Route A proceeds through an epoxide intermediate. Route B involves formation of a five-membered heterocycle containing C, O, and Mn. A is the *trans* diol; B is the *cis* diol.

54. The peroxybenzoic acid OH group reacts with the double bond to form an epoxide and benzoic acid. The resulting expoxide, X, is susceptible to nucleophilic attack and ring opening. Equilibrium protonation of the epoxide oxygen, followed by backside attack of water and concomitant ring opening yields the trans-diol product.

55. Ozonolysis of double bonds yields aldehydes and ketones (after reduction of the intermediate ozonides, which are explosive and thus not isolated). Isolation of formaldehyde means that a terminal CH_2 olefin was present. This must have been bonded to the cyclopentanone ketonic carbon, that is, X must have been a methylene cyclopentane. The two aldehyde functions must have been in a symmetrical six-membered ring (the formula of X is $C_{10}H_{14}$ indicating two rings and two unsaturations).

56. Heterolysis of the weak peroxide bond forms two RO• (radicals). These react with HBr to produce ROH and bromine radicals. The bromine radical attacks the double bond to give the most *stable* secondary organic radical intermediate (putting bromine on the terminal carbon atom). This radical intermediate abstracts hydrogen from HBr

82   VERBAL LEADS

to yield the primary 1-bromobutane product. This regenerates a bromine radical, which reenters the chain process by reacting with more olefin.

57. Sodium fusion shows the presence of a C—Br bond. Sodium ethoxide (NaOEt) treatment eliminates HBr with formation of a new double bond. The initial compound has one double bond (Br_2 and H_2 reactions). This must be a methylcyclopentene because total hydrogenation (uptake of 2 moles of H_2) of the dehydrobrominated product yields methylcyclopentane. This is confirmed by the ozonolysis that produces a *single* keto aldehyde. The double bond must be on a carbon atom to which a methyl group is bonded in order for ozonolysis to produce the single keto–aldehyde. The ketone of the keto–aldehyde must also have the partial structure —$COCH_3$ because it gives a positive iodoform reaction. The original bromide must be allylic for it rapidly precipitates AgBr on treatment with $AgNO_3$, whereas hydrogenation makes this precipitate form less rapidly. The only structure meeting all of these structural requirements is 3-bromo-1-methylcyclopentene. This has the correct molecular weight of 161.

58. Addition of HBr in the presence of peroxides is a free radical addition process in which the Br radical adds to the double bond initially. This addition occurs to give the most stable carbon radical intermediate (tertiary > secondary > primary). Thus the product has hydrogen at the ring fusion and bromine adjacent to the methyl group.

59. A chloronium ion and chloride ion are formed by reaction of the olefin and Cl_2. Backside attack by water on the chloronium ion at the more substituted carbon atom (which can be more carbocation-like) leads to the chlorohydrin product. This product is chiral, but both enantiomers are formed in a racemic mixture because the initial chloronium ion is formed as a racemic pair of ions.

60. This is a typical catalytic reduction of a ketone to an alcohol. The hydrogen can attack the carbonyl group from either side in a cis fashion (i.e., both H atoms of the H_2 add to the same side of the carbonyl group). Attack from one side yields a chiral alcohol with an (*R*) configuration. Attack from the other side yields an alcohol with an (*S*) configuration. Either type of attack is equally probable and so a 50/50 (racemic) mixture of (*R*) and (*S*) products is obtained.

61. Convert the Fischer projection formula to a three-dimensional representation by darkening the horizontal bonds (sticking out of the plane of the page) and dashing the vertical bonds (projecting behind the plane of the page). Assign priorities to the groups attached to C* with H being the lowest priority. View the molecule from the side exactly opposite the group of lowest priority (H) and trace a path from the group of highest priority to the group of lowest priority. The path is clockwise. The configuration is (*R*).

62. The product contains no I, Zn, or Cu. The reaction proceeds through a carbenelike species called a carbenoid. Addition of the carbenoid occurs at each double bond resulting in a product with a total of five rings.

63. The oxygen–oxygen single bond in the peroxide is very weak and easily homolyzes to yield two peroxyacyl radicals $RCO_2 \cdot$. These peroxyacyl radicals lose CO_2. The $R \cdot$ generated functions as an initiator of the polymerization process, adding to the double bond of the *p*-chlorostyrene to give a chain carrying benzylic radical. The benzylic radical adds to another equivalent of p-chlorostyrene to give another larger benzylic radical, and the polymer chain continues to grow in this way. The growing polymer chain can be terminated by coupling with any radical species in the reacting system.

64. Three major types of polymerization of olefinic substrates involve anionic, cationic, and free radical polymerization processes. Anionic polymerization involves nucleophilic attack on the double bond, that is, at the carbon atom bearing the R groups. This is facilitated if R is electron withdrawing and not too bulky. The carbanion produced attacks another molecule of olefin to continue the polymerization process. Cationic polymerization involves electrophilic attack on the olefin, that is, at the carbon atom bearing the R' groups, facilitated if R' is electron donating or stabilizing. The carbocation produced attacks more olefin in a similar fashion to eventually give a polymer. Free radical polymerization occurs in a fashion similar to the anionic and cationic processes, except the intermediate chain carrying species is a carbon radical.

65. Irradiation of diazomethane produces singlet carbene, $:CH_2$. In the liquid phase this will react immediately by inserting into the carbon–carbon double bond. The product with *cis*-2-pentene is thus *cis*-1-ethyl-2-methylcyclopropane. In the gas phase singlet carbene has time to decay to the more stable triplet state. Triplet carbene adds to the double bond of trans-2-pentene in a fashion similar to free radical addition. The adduct has a lifetime long enough for rotation around single bonds and therefore both cis and trans ethylmethylcyclopropane are formed.

66. This reaction must be an example of singlet carbene insertion into single C–H bonds. There are four different types of C–H bonds in the 2-methylbutane substrate. Singlet carbene is so reactive it will react with these essentially in a statistical fashion, inserting between the carbon and hydrogen atoms in each different C–H bond. The number of hydrogen atoms of each type will determine the quantity of each product obtained. The products are 3-methylpentane, 2,2-dimethylbutane, 2,3-dimethylbutane, and 2-methylpentane in ratios of approximately 6:1:2:3, respectively.

67. The first step in this sequence is a free radical chlorination of cyclopentane to yield chlorocyclopentane. The second step is a typical E2 elimination of HCl from the

chlororcyclopentane induced by potassium *t*-butoxide to yield cyclopentene, A. The final step involves reaction of the cyclopentene with dichlorocarbene. This carbene is generated by reaction of chloroform with *t*-butoxide to yield $CCl_3^{\ominus}$, which loses $Cl^{\ominus}$. Insertion of :CCl_2 into the cyclopentene double bond yields the final product, B.

68. Hydroboration of 2-butyne yields a vinylic borane which, when treated with acetic acid at 0°C, is converted to the corresponding cis olefin. Treatment of this cis olefin with peroxyacetic acid yields the corresponding *cis*-dimethylepoxide. Treatment of 2-butyne with ethylamine and lithium at −78°C yields the trans olefin. When this is epoxidized with peroxyacetic acid, it is converted to the corresponding *trans*-dimethyl epoxide.

69. Mercuric ion probably complexes with the triple bond, increasing its susceptability to nucleophilic attack by water. Attack by water leads to an intermediate enol. The oxygen of the enol is bonded to C-2 of the original alkyne so the addition follows Markovnikov's rule. Tautomerization of the enol yields the final product, 2-butanone.

70. It is best to consider such a synthesis by taking apart the product, that is, by approaching it in a retrograde fashion. The cis-olefin product could be obtained by catalytic hydrogenation of the corresponding alkyne with a nickel boride catalyst. The alkyne could be prepared by reacting the sodium salt of 3-phenylpropyne with methyl bromide (S_N2 displacement) and the benzyl alkyne itself could be prepared in a similar fashion from acetylene and benzyl bromide. Sodium amide is the base used to generate these alkyne anions. Benzyl bromide can be prepared from toluene and *N*-bromosuccinimide or bromine. Acetylene results from reaction of calcium carbide with water.

71. Addition occurs in the Markovnikov fashion. The hydrogen of the first equivalent of HCl goes to the carbon atom already bearing the largest number of hydrogen atoms, that is, the terminal carbon atom, thus giving the first intermediate, 2-chloro-1-pentene. The second equivalent of HCl adds in the same way to yield the final product, 2,2-dichloropentane.

72. Treatment of 1-butene with Br_2 in CCl_4 would give a dibromide of formula $C_4H_8Br_2$. Since the next two steps appear to be sequential dehydrobrominations, it is likely that $C_4H_8Br_2$ is the expected 1,2-dibromobutane. Dehydrobromination of this with KOH in ethanol will yield a mixture of the isomeric 1-bromobutene and 2-bromobutene, B and C. Further dehydrobromination with stronger base at higher temperature will yield 1-butyne, D. Treatment of 1-butyne with $NaNH_2$ in liquid ammonia will generate the corresponding acetylide, which will attack the primary carbon atom bearing bromine in ethyl bromide in an S_N2 displacement yielding the symmetrical 3-hexyne. This alkyne will react with diborane, E, at 0°C, to give the corresponding vinylic borane. Alkaline hydrolysis, F, will yield an enol, G, which tautomerizes to 3-hexanone.

73. The substrate is a tertiary bromide and is not susceptible to backside nucleophilic attack. The sodium acetylide is a very strong base and can effectively abstract protons from carbon under appropriate circumstances. Such a situation is ideally suited for an E2 elimination. There are two possible E2 elimination routes. Elimination produces the two olefinic products 2-methyl-1-butene and 2-methyl-2-butene.

74. Hydrogen bromide adds to the diene to give a delocalized allylic carbonium ion. Most of the charge in this cation resides on the secondary carbon atom (C-2) rather than on the terminal carbon atom, and bromide thus adds rapidly at this more positive site to give 3-bromo-1-butene. However, at high temprature this reaction is reversible and the delocalized cation is in equilibrium with the 3-bromo-1-butene product. Addition of bromide can then occur at the terminal carbon atom to give the more thermodynamically stable internal olefin, 1-bromo-2-butene. While this olefin forms less rapidly, it will be the major product under conditions of thermodynamic equilibrium.

75. Heating 2-bromosuccinic acid dimethylester in base will yield the elimination products formed by loss of HBr. The elimination products are dimethyl maleate and dimethyl fumarate. These olefins are good dienophiles in Diels–Alder cycloaddition reactions. They react with cyclohexadiene in a stereospecific fashion, the cis diester giving an endo, cis cyclic adduct and the trans diester giving a trans cyclic adduct.

76. Inscribing cyclobutadiene, cyclopentadienyl anion, and cyclooctatetraene carbon frameworks inside a circle gives an MO picture of the pi systems for each of these species. The four-, five-, and eight-membered ring systems have four, five, and eight MO's, respectively. Cyclobutadiene has one bonding MO, two nonbonding MO's, and one antibonding MO. Cyclopentadienyl anion has three bonding MO's and two antibonding MO's. Cyclooctatetraene has three bonding, two nonbonding, and three antibonding MO's. Filling the MO's according to Hund's rule shows that cyclobutandiene is predicted to be an unstable diradical, that cyclopentadienyl anion is stable (and thus its precursor acid is relatively acidic), and that planar cyclooctatetraene is also not aromatic and therefore not stable as a planar delocalized structure.

77. The best first step in this sequence would be the anti-Markovnikov addition of HBr, in the presence of peroxides, across the double bond of the olefinic substrate. Homolytic cleavage of the R–O–O–R bond generates alkoxy radicals that can abstract a hydrogen atom from HBr to generate bromine radicals. These can add to the double bond to give the most stable benzylic radical. The latter can abstract another hydrogen

atom from HBr, generating 2-bromo-4-phenylbutane and more bromine radicals to carry on the addition process. Reaction of magnesium with 2-bromo-4-phenylbutane yields the corresponding Grignard reagent. Addition of D_2O to this Grignard reagent produces the desired deuterated hydrocarbon along with Mg(OD)Br.

78. The first step is simply a free radical bromination of the benzylic position to produce 2-bromo-2-phenylpropane. This is converted to the corresponding Grignard reagent, X. The Grignard reagent adds to the carbonyl of ethanal to give the *tert*-alcoholate −MgBr adduct, which upon acid work-up yields 3-methyl-3-phenyl-2-butanol. Under acidic conditions this readily protonates on the hydroxyl function, water departs, and a methyl shift occurs to give a stable tertiary benzylic carbocation. This loses a proton to yield A, 2-methyl-3-phenyl-2-butene.

79. Aluminum chloride reacts with 2-chloropropane to produce equilibrium amounts of the corresponding 2-propyl cation and $AlCl_4^{\ominus}$. The 2-propyl cation reacts at the para position of toluene to give a cationic sigma complex in which the 2-propyl carbon becomes bonded to the aromatic ring. This complex loses a hydrogen ion to yield the final product, *p*-isopropyltoluene.

80. Methyl and bromine substituents are *ortho–para* directors; the carboxyl group is a *meta* director. Oxidation of the methyl group to a carboxylic acid group converts an *ortho–para* director to a *meta* director. Reactions sequence (c) ends with the bromination of benzoic acid.

81. 4-Chloroaniline is a Lewis base and aluminum chloride is a Lewis acid. The lone pair on the nitrogen of the amino group reacts with the electron deficient aluminum center in aluminum chloride to form a polar covalently bonded complex. The aluminum in this complex has a formal negative charge and the nitrogen has a formal positive charge. The positive charge on the nitrogen of the amino group strongly deactivates the aromatic ring towards electrophilic attack (i.e., the Friedel–Crafts reaction).

82. The nitro group is an electron-withdrawing substituent by virtue of both resonance and inductive effects. Both the ortho and para positions of nitrobenzene have a small partial positive character in the ground state, which makes electrophilic attack at these sites unfavorable relative to attack at the meta position. The sigma complex formed by attack at the para position can result in a positive charge directly on the carbon atom bonded to the nitro group. The sigma complex formed by attack at the meta position results in a positive charge on carbon atoms not bonded to the nitro group. Since the nitrogen in a nitro group is positively charged, meta attack is favored. *m*-Dinitrobenzene is the product.

83. Both substituents on the benzene ring are *ortho–para* directing. Substitution occurs at both ortho and para positions by similar mechanisms. The aromatic substituent is closer to the reaction center in the ortho substitution process. Steric hindrance between the incoming electrophile and the aromatic substituent during ortho substitution increases the energy of the transition state for this process (relative to para substitution) and thus less ortho product is formed.

84. Both the chloro and amino substituents are *ortho–para* directors by virtue of their electron donating resonance interaction with the benzene ring. The lone pair, which becomes delocalized into the benzene ring, exists on a second row element (nitrogen) when the amino group acts as an *ortho–para* director and on a third row element (chlorine) when the chloro substituent acts as an *ortho–para* director. Orbital overlap from the nitrogen lone pair to the ring is more effective than that of chlorine, as lone pair electrons on chlorine are more diffuse (larger). The major products would be 2-chloro-4-bromoaniline and 2-chloro-6-bromoaniline.

85. Isobutyl chloride is a primary chloride. It complexes with $AlCl_3$ to generate the group $-\overset{\oplus}{Cl}-AlCl_3^{\ominus}$. When this group departs as $AlCl_4^{\ominus}$ it generates positive charge on the carbon atom to which it is bonded. This positive charge (incipient primary carbocation) induces the migration of an adjacent methyl group or hydrogen (as hydride). The resultant secondary or tertiary carbocation reacts with benzene in a typical electrophilic aromatic substitution to give 2-phenylbutane or *t*-butylbenzene, respectively.

86. The methyl group of toluene is an *ortho–para* directing substituent. It is also an activating group. Nitration of toluene with a mixture of concentrated nitric and sulfuric acids will yield a mixture of ortho and para nitrotoluenes. With *very strong* fuming H_2SO_4 and fuming nitric acid, nitration of toluene (or of nitrotoluenes) at high temperatures will yield 2,4,6-trinitrotoluene, TNT. Oxidation of TNT with alkaline potassium permanganate will yield 2,4,6-trinitrobenzoic acid, which will decarboxylate when heated under acidic conditions, to yield 1,3,5-trinitrobenzene. The reason this decarboxylation is so facile is that the incipient negative charge at the transition state for the reaction is stabilized inductively by the trinitro functionality of the ring. This stabilization is not a resonance interaction, as the negative charge is orthogonal to the pi system of the ring.

87. The substrate in this reaction has three strong electron-withdrawing substituents bonded to a carbon atom bearing a single hydrogen atom. Methylmagnesium bromide is a highly polar species and the methyl moiety has partial methide anion character (i.e., it can act as a base). The methyl group of methylmagnesium bromide abstracts the acidic hydrogen of the substrate to yield methane (which bubbles out of the mixture) and $[CH_3COC(CN)_2]^{\ominus} \cdots \overset{\oplus}{Mg}Br$.

88. The reaction does not go in pure methylene chloride because sodium cyanide is not soluble in methylene chloride and because 1-bromobutane is not soluble in water. These phases remain separate. The crude methylene chloride was probably contaminated with triethylamine, which reacted with 1-bromobutane to produce small amounts of butyltriethylammonium bromide. The quaternary ammonium compound has a partial solubility in both water and methylene chloride, along with its gegenion. This latter ion can be either bromide or cyanide since both are present in the mixture. Thus a quaternary ammonium cyanide can have partial solubility in the methylene chloride and this provides a means for bringing cyanide ion and 1-bromobutane into the same phase so reaction can occur. The quaternary compound is acting as a phase-transfer agent.

89. The alcohol function of the substrate is about as strong an acid as methanol so that there can be an equilibrium deprotonation of the OH function in the side chain by methoxide. The resulting alkoxide function is strategically located adjacent to the chloride that is activated for nucleophilic aromatic substitution. An internal displacement of chloride ion can occur to yield the benzodihydrofuran product.

90. Friedel–Crafts alkylation and acylation are starting points for these preparations. Using an aluminum chloride catalyst and chloroethane as the reagent, benzene can be converted to ethylbenzene. Benzylic bromination of ethylbenzene (i.e., with bromine and light) will yield 1-bromo-1-phenylethane. This can be hydrolyzed in water to the desired 1-phenylethanol. Acylation of benzene using ethanoyl chloride and one equivalent of aluminum chloride yields an acetophenone-aluminum chloride complex, which can be decomposed in water to yield acetophenone. This ketone can be reduced with lithium aluminum hydride or sodium borohydride to the desired 1-phenylethanol.

91. Reaction of metallic sodium with alcohols produces the corresponding alkoxide ion, sodium ion, and hydrogen gas. Ethers do not react at all. The most stable alkoxide is also the most easily formed. Alkyl groups with their electron-releasing tendencies exert a destabilizing effect on the corresponding negatively charged alkoxide ions (products). The primary alkoxide from (b) is the most stable and the most rapidly formed.

92. Consider the use of a Grignard reagent to lengthen the carbon chain. 2-Propanol is converted to 2-chloropropane with PCl_3. The 2-chloropropane reacts with magnesium under anhydrous conditions to produce the Grignard. The Grignard reagent reacts with ethylene oxide to form 3-methyl-1-butanol after acid work-up.

93. The product, an ester, can be prepared from the reaction of an acid with an alcohol. The acid is prepared by oxidizing acetaldehyde with dichromate or other appropriate oxidizing agents. The alcohol is prepared by reduction of acetaldehyde with lithium aluminum hydride or other appropriate reducing agents.

94. The two charges in compound A repel each other and neither can be easily delocalized into the ring without approaching the other. Alkyl groups inductively destabilize carbanions so that compound C is less stable than compound B. The phenolic anion–carbanion interaction in compound A is much more effective at destabilizing this species than the methyl group destabilizations in compound C.

95. Because chloride ion is a weak nucleophile it is unable to displace the group $-OH_2^{\oplus}$ formed by protonation of this primary alcohol. Zinc reacts with the hydrochloride acid in the reaction mixture to produce zinc chloride and hydrogen gas. Zinc chloride coordinates with the nonbonding electrons on the oxygen atom of the alcohol to produce the better leaving group $-\overset{\overset{\oplus}{|}}{\underset{H}{O}}-\overset{\ominus}{ZnCl_2}$, which is more easily displaced by chloride ion.

96. The first step in this sequence, regardless of whether or not triethylammonium chloride is present, is the formation of the corresponding 2-propyl chlorosulfite by attack of the hydroxyl on the sulfur of thionyl chloride. The chlorosulfite can lose sulfur dioxide to give an ion pair, which can rapidly collapse to give a chloride of retained configuration. In the presence of chloride (anion) from the triethylammonium chloride, however, some backside attack on the chlorosulfite intermediate can occur giving a chloride of inverted configuration.

97. Consider the polarization of the Grignard reagent. Carbon next to magnesium is electronegative; magnesium is electropositive. The electronegative carbon atom of the Grignard attacks the electropositive carbon atoms of the other reactants in reactions (a)–(d); oxygen associates with magnesium. Two moles of Griganrd are needed for complete reaction with the ester in reaction (d). Acid work-up of the magnesium salt produces the alcohol.

98. Reaction (1) involves reaction of a chiral tertiary substrate in an ionizing solvent, whereas reaction (2) involves reaction of a chiral primary substrate under nucleophilic reaction conditions. Reaction (1) is a typical $S_N 1$ displacement, whereas reaction (2) is a typical $S_N 2$ displacement. Reaction (1) proceeds through a tertiary carbocation that gives a racemic product. Reaction (2) proceeds with inversion to give a product with inverted configuration.

99. The secondary bromide in Compound A will ionize readily to give a secondary carbocation. This type of ionization is aided dramatically by the nonbonding electrons on the oxygen of compound B, which provide a resonance form where the positive charge can reside on oxygen. The strongly electronegative fluorines make the CF_3CF_2 group in compound C electron withdrawing, thus destabilizing the incipient carbocation that forms as bromine departs. Thus Compound C will ionize least rapidly.

100. In such a reaction, the choice of the nucleophile is critical. Either methoxide or t-butoxide must be used. What side reaction can occur under the conditions chosen by the student? Using a strong base (sodium methoxide) in the presence of t-butyl bromide results in elimination exclusively. Better choices for this synthesis would have been methylbromide and t-butoxide.

101. Protonation of oxygen provides two alternative routes for C—O bond cleavage. Which of the covalent bonds to oxygen should be broken? Heterolytic cleavage to produce t-butyl alcohol will result in the unstable phenyl carbocation as an intermediate. Alternatively, cleavage of the other bond to oxygen produces phenol and the relatively stable t-butyl carbocation.

102. What is the function of methyllithium? The methide ion of methyllithium abstracts the hydroxylic proton from the substituted cyclopentanol to produce the corresponding alkoxide ion. Nucleophilic attack by this alkoxide on sulfur yields the tosylate ester.

103. In this Williamson synthesis either phenoxide ion or methoxide ion must be used as the nucleophilic agent. An alkyl iodide serves as the substrate for the nucleophile. The choice of base is critical. Which base is better for this purpose, phenoxide or methoxide ion? Why? Nucleophilic displacement of the iodide from methyl iodide can produce the anisole. The alternative, displacement of iodide by methoxide ion from an iodobenzene, is not attractive because this requires displacement of $I^{\ominus}$ from the benzene ring.

104. This is an example of the Reimer—Tiemann reaction. Strong base added to $HCCl_3$ will generate dichlorocarbene. Nucleophilic attack by phenoxide on electrophilic dichlorocarbene occurs. Tautomerization follows to regenerate the aromatic system. Displacement of one chloride on the dichloro intermediate ion by hydroxide ion is followed by loss of HCl to produce the aldehyde.

105. This is an example of the pinacol rearrangement. The reaction begins by protonation of oxygen. The protonated alcohol can lose water and ring expand. Ring expansion occurs after a tertiary carbocation is formed. Finally, proton loss produces the desired spiroketone.

106. The function of the peracid is to expoxidize the olefin. The epoxide intermediate undergoes ring opening in the presence of additional m-chloroperbenzoic acid to produce an intermediate with both ester (lactone) and alcohol functionality. The macrocyclic lactone product is formed when m-chlorobenzoic acid is lost from this intermediate.

VERBAL LEADS 91

107. Give careful consideration to drawing the correct structural formulas. Protonation of the alcohol by sulfuric acid is followed by loss of water to produce a tertiary carbocation. Ring expansion produces the substituted cyclopentyl cation. A methide shift, followed by proton loss, yields the Saytzeff product.

108. The inductive effect of the *meta t*-butyl groups is not significant. Why? Consider stabilization of the conjugate base as a criterion for acidity. Once generated, the phenoxide ion from compound A is more stable than the phenoxide ion from compound B. Why? The steric effect of the *t*-butyl groups forces the nitro group out of the ring plane.

109. Formation of a chromate ester occurs via nucleophilic attack by the oxygen of the alcohol on the electropositive chromium. A proton is lost from the carbon atom adjacent to the chromate ester. At the same time loss of $HCrO_3^\ominus$ yields acetone.

110. Protonation of the oxygen makes the carbonyl carbon atom more susceptible to nucleophilic attack. The nonbonding electrons on oxygen in methanol attack the carbonyl carbon atom of the ketone. A proton is lost to form the hemiacetal.

111. The key here lies in understanding how diazomethane is polarized. A resonance structure with negative charge residing on carbon is a principal contributor to the ground state of diazomethane. This carbon atom is therefore nucleophilic. The carbonyl carbon atom is attacked by diazomethane. Nitrogen (N_2) is lost, the ring expands, and the carbonyl is regenerated.

112. The cyanide ion is a strong nucleophile. (HCN is a fairly weak acid.) The cyanide ion attacks the carbonyl carbon atom. Protonation of the addition product results in cyanohydrin A. Hydrolysis of cyanohydrin A produces acid B.

113. Nucleophilic attack on the carbonyl carbon atom by nitrogen's free electron pair starts this reaction. Protonation of the electronegative oxygen is followed by another proton transfer. This product is now ready for dehydration. Loss of water is accomplished via the internal attack on carbon by the electron pair from nitrogen. Loss of a proton yields the oxime.

114. Each reaction begins with an acid catalyzed nucleophilic attack on the carbonyl carbon atom of cyclohexanone. The free electron pair on nitrogen ($-\ddot{N}H_2$) acts as the nucleophile. The adduct, resulting from the nucleophilic attack, dehydrates to form the Schiff's base, C=N, in all four reactions.

115. Ketal formation typically begins with the (acid catalyzed) formation of a hemiketal. Water is lost from the hemiketal under acidic conditions to give a positively charged intermediate. Internal nucleophilic attack (by the second oxygen of the ethylene glycol) forms the ketal.

116. The most acidic proton is on the nitrogen of the hydrazone. This proton is transferred to the carbon via base catalyzed tautomerization. The second proton is lost from nitrogen, followed by eventual loss of N_2. The resulting carbanion is protonated to give the hydrocarbon product.

117. What is a common use for ethylene glycol? Ketal formation via the action of ethylene glycol involves only ketone functionality. Esters can be reduced by $LiAlH_4$. Regeneration of the ketone from the ketal allows the reduction of only the ester.

118. What is the function of Cu_2Cl_2? Association of copper ion with the carbonyl oxygen atom renders the β-carbon atom more susceptible to attack by the electronegative moiety of the Grignard reagent. Attack occurs via a Michael type addition. Without cuprous chloride the reaction is simply a normal Grignard addition to a carbonyl.

119. There is an acidic proton on nitromethane. Abstraction of this proton by ethoxide ion produces the conjugate base, which can act as a nucleophile and attack the carbonyl carbon atom. Protonation of the sodium salt yields the nitromethane adduct (**III**). Reduction of the nitro group yields an amine alcohol. Diazotization with HONO follows. Loss of nitrogen and ring expansion yields the product.

120. This is an example of a Wittig reaction. The strongly basic tertiary butoxide abstracts a proton from the triphenylphosphonium bromide salt to produce the ylide. The resulting ylide attacks the carbonyl carbon atom and forms the betaine. The betaine subsequently loses triphenylphosphine oxide to produce the olefin.

121. This is an example of the Baeyer–Villiger oxidation. Reaction begins with a nucleophilic attack on the carbonyl carbon atom and protonation of the carbonyl oxygen atom. The resulting adduct loses trifluoroacetic acid after undergoing an additional protonation from the acid (solvent). The loss of trifluoroacetic acid generates an electron deficient oxygen ready to accept the migrating group. The *p*-nitrophenyl group does not migrate.

122. With no acidic alpha protons the only reasonable reaction path is hydroxide attack on the carbonyl functionality. After hydroxide ion adds to the carbonyl carbon atom, contraction to the fluorene ring system occurs with concurrent formation of carboxylic acid functionality. A proton shift follows to give the α-hydroxy acid, a typical benzilic acid rearrangement product.

123. The Reformatsky reaction extends the carbon skeleton of an aldehyde or a ketone and yields the β-hydroxyester. The choice of starting materials should include

$$CH_3-\underset{\underset{CH_3}{|}}{CH}-CH_2-\underset{}{\overset{\overset{H}{|}}{C}}=O, \quad Br-\underset{\underset{CH_3}{|}}{\overset{\overset{CH_3}{|}}{C}}-C\underset{OEt}{\overset{\nearrow O}{\diagdown}}$$

zinc, and $H_3O^{\oplus}$. The mechanism is similar to that of the Grignard reaction. Zinc complexes with the α-bromo ester to produce the organozinc reagent. The zinc and the carbon of the organozinc reagent combine with the oxygen and the carbonyl carbon atom of the aldehyde, respectively. Hydrolysis produces the desired hydroxyester.

124. These reactions are all base-catalyzed aldol condensations. The starting carbonyl containing compound produces an enolate anion in basic solution. The enolate anion attacks the carbonyl carbon atom of the other reactant to produce an adduct. Upon work-up, the adduct yields the β-hydroxycarbonyl compound.

125. What is the function of the hydride ion? Where are the potential reactive sites on acetone? Hydride acts as a base by removing the acidic α proton on acetone. Displacement of iodide by the resultant enolate follows.

126. Cyanide ion attacks the carbonyl carbon atom to begin this reaction, known as the benzoin condensation. The resulting anion undergoes a 1,2-proton shift generating the α-anion of 1-hydroxyphenylacetonitrile. This anion attacks the carbonyl carbon atom of another molecule of benzaldehyde. The resulting adduct undergoes an intramolecular proton transfer, followed by expulsion of cyanide.

127. This is a typical internal adol cyclization. The first step involves proton loss to form a ketolate anion. An aldehyde is more susceptible to nucleophilic attack than a ketone because the carbonyl carbon atom of the aldehyde is more electropositive. (This is due to the electron-releasing inductive effect of the alkyl groups in the ketone.) The resulting cyclic intermediate then undergoes protonation followed by dehydration to produce the unsaturation in conjugation with the carbonyl.

128. This is an aldol condensation product. The aldol condensation is reversible. In a strongly basic solution this compound loses the hydroxylic proton. Internal nucleophilic displacement on carbon occurs to form acetone and the conjugate base of acetone. Protonation subsequently produces the completed retro-aldol condensation products.

129. The effect of the substituted benzene ring is to make the carbonyl carbon atom of the aromatic aldehyde less electropositive than the carbonyl of formaldehyde. The first

step involves attack of hydroxide ion on the carbonyl carbon atom of formaldehyde. In the second step, oxidation–reduction takes place through a hydride transfer to the m-methoxybenzaldehyde. This addition product is protonated by the formic acid to produce the product.

130. The most acidic proton is alpha to the ketone. Loss of this proton produces a carbanion that attacks the carbonyl function of the ester. This forms a negatively charged bicyclic intermediate. This bicyclic intermediate opens to form an alkoxide ion that attacks the carbonyl of the substituted cyclohexanone ring to form a γ-lactone. Internal attack by the carbanion alpha to the ketone on the γ carbon of the lactone produces the cyclopropane ring and the salt of the acid. Protonation yields the desired product.

131. This is the iodoform reaction. The base removes the acidic α-hydrogen and the resulting carbanion attacks iodine. This occurs two more times until all the α-hydrogen atoms have been replaced by iodine. (Introduction of the first iodine makes the remaining two hydrogen atoms more acidic.) After all the α-hydrogen atoms have been replaced by iodine, the carbonyl carbon atom of $\phi COCl_3$ undergoes nucleophilic attack by hydroxide ion.

132. Consider a base catalyzed aldol condensation followed by dehydration. The singlet at $\delta 2.14$ indicates a possible vinyl methyl group. Absorption at 1702 cm^{-1} suggests a carbonyl. Inspection of the NMR data indicates that only two protons can exist α to the carbonyl in the final product. Thus, cyclization can occur in only one direction.

133. Consider the acidity of the α protons. Loss of an α proton to methoxide ion is followed by internal nucleophilic displacement of chloride ion. This results in formation of a strained bicyclic ketone intermediate. Ring opening of this bicyclic ketone occurs via methoxide attack on the carbonyl carbon atom. Protonation of the resulting carbanion by methanol yields the product. This is an example of the Favorski rearrangement.

134. This is an example of the Hell–Volhard–Zelinsky reaction. A is an α-bromocarboxylic acid. Nucleophilic displacement of bromide produces the amino acid alanine.

135. Consider the reactivity of the carbonyl carbon atom. The presence of chlorine on the carbonyl carbon atom makes it particularly susceptible to nucleophilic attack. The carbon–chlorine bond breaks because the alternative, cleavage of the carbon–carbon bond, requires more energy. The highly reactive acid chlorides produce an amide, an ester, an amide, and an anhydride, in reactions (a)–(d), respectively.

136. Consider the polarization and reactivity of the carbonyl carbon atoms in each of these carboxylic acid derivatives. The electropositive carbonyl carbon atom undergoes a nucleophilic attack by a nonbonding electron pair of water, producing an intermediate adduct. Loss of an appropriate leaving group from the adduct produces acetic acid.

137. Phosgene (ClCOCl) and methyl amine (CH_3NH_2) produce methyl isocyanate. Consider another route. Phosgene can react with α-naphthol to produce α-naphthyl chloroformate. The chloroformate can react further with methylamine to produce Carbaryl.

138. Consider the function of the acid catalyst. Protonation of the carbonyl oxygen atom of the carboxylic acid by the acid catalyst renders the carbonyl carbon atom of the acid more susceptible to nucleophilic attack. Ring closure occurs via nucleophilic attack on the carbonyl carbon atom by the oxygen of the hydroxyl group. The resulting positively charged cyclic intermediate undergoes a proton shift before losing water to form the lactone.

139. Acetic acid is an ionizing solvent in which S_N1 type solvolysis can occur. Loss of tosylate is accompanied by a 1,2-methide shift. Proton loss from the resulting carbocation generates the hydrocarbon product, 1,1,2,3-tetramethylindene.

140. Formation of $C_6H_5CH_2CO_2Br$ is followed by fragmentation to the corresponding carboxy radical and bromine atom. The carboxy radical loses CO_2 to form a benzyl radical. This abstracts a bromine atom from another $C_6H_5CH_2CO_2Br$ to form benzyl bromide. The process is repeated.

141. This esterification is acid catalyzed. Which oxygen has the greatest ability to accomodate a proton? Protonation of the carbonyl oxygen atom leads to an electropositive carbon that is susceptible to nucleophilic attack by the labeled oxygen. The adduct formed undergoes proton transfer to either of the other geminal oxygen atoms. The newly formed positive species loses water and then a proton to form the labeled ester.

142. Aspirin is acetylsalicylic acid. The reaction taking place is a simple acid catalyzed hydrolysis of an ester beginning with protonation of the carbonyl oxygen atom. Nucleophilic attack by water on the carbonyl carbon atom is followed by proton transfer and formation of salicylic acid and acetic acid.

143. Protonation of nitrogen renders the carbon atom of the nitrile more vulnerable to nucleophilic attack by the electron rich oxygen atom of water. The resulting intermediate undergoes a proton shift and then a second water molecule attacks as in the previous step to give another intermediate. Loss of ammonia then results in benzoic acid.

144. Consider the resonance and inductive effects of the —OH group. These effects oppose each other since —OH withdraws electrons inductively and donates electrons via resonance. Electron donors tend to stabilize the protonated form rather than the conjugate base, whereas electron-withdrawing groups tend to stabilize the conjugate base. In this problem these effects can be sorted out.

145. Consider the acidities of the three molecules. Benzoic acid is a fairly good acid (pK_a 4.5), phenol is a weaker acid (pK_a 10), and biphenyl does not dissociate at all. Washing the ether solution with 5% $NaHCO_3$ removes the benzoic acid and produces sodium benzoate in the aqueous layer. After separating the layers the ether can be washed a second time with stronger base (10% NaOH). This washing removes phenol from the ether as sodium phenoxide. The ether layer is then dried (over $MgSO_4$) and the solvent is removed to yield biphenyl.

146. Loss of the first proton makes hydrogen bonding possible between the hydrogen of an unionized carboxyl group and the oxygen of an ionized carboxylate ion. It is necessary to consider the repulsive forces in the dianion in order to correctly explain the second part of the question. The carboxyl groups are farther apart in terephthalic acid than in phthalic acid.

147. Protonation of oxygen is followed by loss of water. As water departs, the phenyl group migrates to the electron deficient nitrogen generating a carbocation. The carbocation undergoes nucleophilic attack by water. Following a proton loss, tautomerization produces the amide. The group that migrates is anti to the departing —OH.

148. Consider the relative ease of abstraction of the acidic protons and/or relative stabilization of the corresponding conjugate bases. The nitro group withdraws electrons via the inductive and resonance effects. The presence of an electron-withdrawing group increases the acidity. The methyl group is an electron donor.

149. Oxidation of the double bond by permanganate yields *cis*-1,3-cyclohexane dicarboxylic acid. Heating this diacid under anhydrous conditions in the presence of an acid catalyst results in attack by the hydroxyl of one CO_2H group on the carbonyl group of the other CO_2H. Via a series of prototropic transfers, $H_3O^\oplus$ is eventually lost to yield the bridged anhydride.

150. Electrophilic attack on the ring occurs at the ortho position. Attack on oxygen is not productive. Why? Tautomerization yields the substituted phenol. Acidification produces salicylic acid. This is an example of the Kolbe reaction.

151. What are the formal charges on carbon and nitrogen in diazomethane? Recall that FC = $Z - (U + S/2)$ where Z refers to the family in the Periodic Table, U refers to the number of unshared electrons, and S to the number of shared electrons, respectively. The electronegative carbon of diazomethane is protonated by the carboxylic acid. The resulting ion pair is composed of the methyl diazonium ion and the carboxylate ion. Nucleophilic attack by carboxylate ion on the methyl diazonium ion with concomitant loss of N_2 yields the product.

152. Mesitylene undergoes bromination to produce bromomesitylene. (This is an example of a typical electrophilic aromatic substitution.) The bromide can be reacted with magnesium to form the corresponding Grignard reagent. Reaction of the Grignard with CO_2 is followed by an acid work-up.

153. Compound A results from an electrophilic aromatic substitution on benzene. Compound B reacts with methyltriphenylphosphonium bromide via a Wittig reaction to produce compound C. The Simmons–Smith reagent converts C to D.

154. It is best to begin by looking at the product and working backwards. The ester bond is likely to have been formed last by reaction of the corresponding acid chloride and alcohol. The required acid chloride can be obtained by bromination of the starting material with bromine and a ferric bromide catalyst, making the corresponding Grignard reagent from the para-bromide product and adding it to carbon dioxide (dry ice). An acid work-up gives the corresponding carboxylic acid, which is readily converted to the required acid chloride with thionyl chloride. The required alcohol can be obtained from the same Grignard reagent (vide supra) by reacting it with ethylene oxide.

155. Compound A is acidic and compound B is less so. This observation, combined with the fact that compound B is converted into compound A upon standing in air, apparently with incorporation of oxygen (MWA − MW B = 16), leads to the conclusion that compound B may be an aldehyde and compound A the corresponding carboxylic acid. A positive Tollens' test and DNP formation for compound B (but not for compound A) confirms this supposition. Thionyl chloride treatment of compound A would then give the acid chloride C, which under Friedel–Crafts conditions would cyclize to an α-tetralone only if C was a phenylbutanoyl chloride. Two α-tetralones are isolated because there is a methyl group on the benzene moiety of C and two alternative modes of cyclization can occur to yield the two different methyl-α-tetralones.

156. The acidic methylene hydrogen between the carbonyl and the amine is removed. The resulting carbanion is stabilized by the adjacent carbonyl group. This carbanion attacks the carboethoxy carbonyl carbon atom across the ring to form a bicyclic intermediate. This intermediate loses ethoxide ion to produce a bicyclic β-ketoethyl ester. Acid hydrolysis (HCl) of this ester produces the corresponding β-keto acid. Decarboxylation of this β-keto acid is followed by protonation of the amine functionality by HCl to produce the hydrochloride salt.

157. Diazonium salt formation occurs. Loss of nitrogen and concomitant ring expansion yields a stable carbocation. This newly produced carbocation is stabilized by the unshared pair of electrons on the adjacent oxygen. Loss of a proton yields cycloheptanone.

158. The product caused one of the worst chemical disasters in history in Bhopal, India. Hydrochloric acid (2 moles) is also a product of the reaction. The reaction begins with attack on the positive carbonyl carbon atom by the free pair of electrons on nitrogen. Dehydrochlorination follows with concomitant formation of methyl isocyanate.

159. The Sandmeyer reaction begins with the diazotization of aniline with sodium nitrite and HCl at low temperatures. The diazonium salt (product of the Sandmeyer reaction) can be converted to benzonitrile (ϕ−CN) via the action of CuCN. Hydrolysis of the nitrile produces benzoic acid.

160. Consider polarization of the isocyanate. The most electrophilic carbon of CH_3−N=C=O reacts with the nucleophilic oxygen of α-naphthol. A proton transfer produces 1-naphthalenol methylcarbamate (Carbaryl), the active ingredient in the insecticide SEVIN®

161. Acetic anhydride reacts with aniline to produce compound A, which has an amino group. Treatment of compound A with chlorosulfonic acid results in an electrophilic aromatic substitution at the para position producing compound B. Reaction of compound B with 2-aminothiazole produces a compound with both carboxamide and sulfonamide functionality. Carboxamides hydrolyze much more rapidly than sulfonamides. Why?

162. Consider the Hinsberg test. In this test primary amines react with benzenesulfonyl chloride to form *N*-substituted benzenesulfonamides that are soluble in base. Secondary amines react with benzenesulfonyl chloride in aqueous KOH to form insoluble *N,N*-disubstituted sulfonamides. Tertiary amines do react with benzenesulfonyl chloride to form *N,N,N*-trialkyl-*N*-benzenesulfonylammonium chlorides, but these are unstable under basic conditions so they decompose back to the tertiary amine.

163. The electron pair on nitrogen provides the nucleophilic "push" for the reaction of pyrrole with electrophiles at the β position of the pyrrole ring. This electron pair is in a *p* orbital approximately parallel to the pi bond orbital involving the α- and β-carbon

atoms. Electron flow from pyrrolic nitrogen to the β position readily occurs. The nonbonding pair of electrons on the nitrogen in pyridine is in an sp^2 orbital that is orthogonal to the pi system. This electron pair cannot be delocalized to the β (meta) position.

164. These reactions constitute a sequence of nitrogen alkylations with methyl iodide and Hoffmann eliminations with Ag_2O. Product A is the dimethyl quaternary iodide of the starting material. This undergoes elimination and ring opening with Ag_2O to give 2,3-dimethyl-4-dimethylamino-1-butene. The latter butene is alkylated to give compound C, which eliminates trimethylamine to yield 2,3-dimethylbutadiene, D.

165. Diazotization of a primary aliphatic amine results in the formation of an unstable diazonium salt. (This is different from diazotization of a primary *aromatic* amine that produces a stable diazonium salt as in the Sandmeyer reaction.) Loss of nitrogen from the unstable diazonium salt yields a carbocation. Carbocation formation is followed by ring contraction, elimination, or substitution.

166. This is a *nucleophilic* aromatic substitution. On which ring would you expect substitution to occur? Since nitrogen is more electronegative than carbon, nucleophilic substitution takes place on the nitrogen containing ring. Which atom sustains nucleophilic attack? Normally the number 2 position is attacked by nucleophiles, but in this case the attack is blocked. The number 4 position thus becomes the reaction site with attack by amide ion producing a resonance stabilized anionic sigma complex. A hydride is lost from the number 4 position during acid work-up.

167. Compound A must possess carbonyl functionality in order to react with hydroxylamine hydrochloride. The product of this reaction is an oxime. The Beckmann rearrangement converts oximes to amides. Amide hydrolysis yields an acid and an amine upon work-up. The amine is characterized as the acetyl derivative.

168. This is an example of the Gabriel synthesis. The neutralization equivalent is equal to the molecular mass divided by the number of acidic protons on the molecule. The neutralization equivalent suggests phthalic anhydride and phthalic acid as the best choices for compounds A and F, respectively. Formation of potassium phthalimide is followed by nucleophilic displacement of bromide from butyl bromide by the phthalimide. Hydrolysis of the resulting substituted phthalimide produces butylamine.

169. In electrophilic aromatic substitution, pyridine is less reactive than benzene, but in *nucleophilic* aromatic substitution, pyridine is *more* reactive than benzene. This is because the more electronegative nitrogen atom in pyridine is better able to accommodate negative charge that must be stabilized during nucleophilic substitution. Nucleophilic attack occurs in the two position leading to a resonance stabilized intermediate with substantial negative charge residing on nitrogen. Loss of hydride regenerates the aromatic ring.

170. The first step involves formation of an *N*-bromoamide. (This is a base catalyzed bromination.) The *N*-bromoamide has an acidic proton due to the electron-withdrawing bromine atom. Loss of this proton produces a nitrene that rearranges to the isocyanate. The isocyanate reacts with two equivalents of base to produce the amine and the carbonate ion.

171. The structure of dicyclohexylcarbodiimide is $C_6H_{11}N=C=N-C_6H_{11}$. The carbon of the carbodiimide undergoes nucleophilic attack by the oxygen of the acid to produce an adduct. This adduct undergoes proton transfer from oxygen to nitrogen, followed by nucleophilic attack of propyl amine on a carbonyl function. Finally, formation of *N,N*-dicyclohexyl urea is accompanied by formation of $C_{10}H_{13}NO$, an amide.

172. A schiff's base is formed from the aldehyde and the amine. The protonated Schiff's base undergoes nucleophilic attack by an electron pair from benzene to produce a positively charged intermediate. The aromatic system is regenerated after proton loss.

173. The most stable conformation for the dicarboxylic acid involves a six-membered ring stabilized by hydrogen bonding. The driving force for the reaction is formation of CO_2. Once carbon dioxide is lost the resulting enol tautomerizes to give the final product.

174. Consider the function of diethylamine. The amine acts as a base by abstracting a proton from the α-carbon of diethyl malonate. The resulting anion acts as a nucleophile. Consider the possible sites on nucleophilic attack on *p*-bromobenzaldehyde. Nucleophilic attack on the carbonyl carbon atom of *p*-bromobenzaldehyde is followed by dehydration to produce the condensation product. This is an example of a Knoevenagel reaction.

175. Hydroxide abstracts the hydroxylic proton of the substrate with concomitant formation of a carbonyl function at this point as the ring opens and the α,β-unsaturated carbonyl in the other ring becomes an enolate. This enolate is nucleophilic. The α,β-unsaturated carbonyl moiety formed from opening of the other ring is electrophilic. The enolate adds to the β carbon of the α,β-unsaturated ketone moiety forming a new eight-membered ring ketone.

176. Consider the function of ethoxide ion. Attack on the carbonyl of the acetate or benzoate ester by ethoxide ion is not productive since it merely results in regenerated

starting material. On the other hand, loss of an α-proton from ethyl acetate produces a useful carbanion. This carbanion displaces ethoxide from the substituted ethyl benzoate to produce the product. This is an example of a crossed Claisen condensation.

177. This is an example of a malonic ester synthesis. The bromides of the dibromopropane are sequentially displaced. (The nucleophilic site in each displacement is between the carbonyl carbon atoms.) The cyclobutane carboxylic acid, D, is produced after hydrolysis of the intermediate diester, B, and subsequent decarboxylation of the resultant diacid, C.

178. Acetoacetic ester loses its central acidic proton to NaOEt producing sodioacetoacetic ester. The sodioacetoacetic ester, acting as a nucleophile, displaces bromide ion from bromoacetone to form compound A. Hydrolysis and neutralization then yield compound B.

179. A reducing sugar causes the reduction of another compound or ion. The reduction of copper ions, $Cu^{2\oplus}$, to produce a brown-red precipitate, Cu_2O, is used to confirm the presence of a reducing sugar. In order to qualify as a reducing sugar, the sugar itself must be capable of undergoing oxidation. Aldehydes meet this criterion. Hemiacetals also qualify as reducing sugars because in aqueous solution, the hemiacetal is in equilibrium with its aldehyde precursor. Glycosidic linkages such as those found in Compounds C and D do not hydrolyze easily.

180. Compounds with hydroxyl groups on adjacent carbon atoms undergo oxidative cleavage when they are treated with aqueous periodic acid. In addition, a carbonyl carbon atom adjacent to a carbon with a hydroxyl group can undergo oxidation. Periodic acid oxidation requires a new C—O bond to be formed for every C—C bond that is broken. Formation of CO_2 indicates the presence of a ketone, in this case a ketose. Since sedoheptulose can be degraded to D-altrose this indicates that the stereochemistry on carbon atoms 2–5 in D-altrose and 3–6 in sedoheptulose is the same. Clearly sedoheptulose is of the D-series.

181. Table sugar (sucrose) is a disaccharide that yields glucose and fructose upon acid catalyzed hydrolysis. Vinegar contains acetic acid. Could the sum of the sweetness factors of the hydrolysis products be more than the sweetness factor for sucrose?

182. This reverse aldol condensation begins with proton abstraction from the —OH on C_4. Heterolytic cleavage of the covalent bond between C_3 and C_4 results in the formation of D-glyceraldehyde, A, and a carbanion. The carbanion transfers a proton from the carbon adjacent to the negative charge to produce an enolate. This enolate subsequently undergoes proton transfer to produce a new enolate, which is the conjugate base of dihydroxyacetone. Protonation results in the formation of dihydroxyacetone.

Combination of D-glyceraldehyde and dihydroxyacetone to form D-fructose is an example of an aldol condensation.

183. Glucaric acid is a 1,6-diacid. An obvious choice is D-glucose, but consider the L-series for other candidates. Could oxidation of carbon atoms 1 and 6 of L-gulose produce the same product as the oxidation of D-glucose?

184. Consider the statistical possibilities resulting from the esterification of glycerol with these three acids. Is there a stereogenic (chiral) center? If so, what effect does it have on your answer? The stereogenic (chiral) center is on the number two carbon atom of glycerol.

185. Consider which peptides have side chain functionality capable of ionization. What are the contributions to the total charge of each peptide by the side chain functionality at pH 1.5? Look up specific pK_a values if necessary. At pH 1.5 Lys (in II) contributes a charge of +1 because of its amino group. The N-terminal amino group is positively charged, but this is true for I, II, and III. At pH 1.5 those peptides least likely to migrate toward the cathode have acidic side chains, that is, Compound I. Compound III has no basic side chains.

186. The function of 2,4-dinitrofluorobenzene (Sanger's reagent) is to react with any free amino group. Which other amino acids in this peptide have potential reaction sites? Ornithine's α-amino group is free to react. Basic hydrolysis produces DNFB–Val as well as DNFB–Orn.

187. Cyanogen bromide is a polar reagent with a partial positive charge on the carbon atom and a partial negative charge on the bromine atom. The sulfur atom on methionine attacks the carbon of cyanogen bromide to form a sulfonium ion intermediate. The loss of CH_3SCN occurs with concomitant ring formation resulting from nucleophilic attack on the α carbon of methionine. The nucleophile is the carbonyl oxygen atom of the methionine peptide bond. The ring-containing intermediate contains $\overset{\oplus}{\diagdown C=NH-}$ functionality, which undergoes hydrolysis to cleave the peptide.

188. Where can you find an electropositive site on phenyl isothiocyanate? Nucleophilic attack by the free pair of electrons on the amino group of the N-terminal amino acid occurs on the electropositive carbon of the isocyanate. This generates an electron rich nitrogen (next to the phenyl group) that can be used for a second nucleophilic attack. Where should this attack occur? Attack by nitrogen takes place on the carbonyl carbon atom of the first peptide bond in the peptide chain. This results in C–N bond cleavage to produce a polypeptide with one less amino acid residue as well as the released phenylthiohydantoin.

VERBAL LEADS 103

189. Consider the function of carbobenzoxy chloride. This reagent protects the amino group from unwanted side reaction by forming carbobenzoxyglycine. The $SOCl_2$ produces the acid chloride of carbobenzoxyglycine thus allowing acylation of the amino group of alanine to produce the peptide bond. Following formation of the peptide bond, hydrogenation yields toluene and glycylalanine (a deprotection reaction).

190. The reaction begins with Schiff's base formation between ninhydrin and the amino group of glycine. This first Schiff's base decarboxylates to produce a second Schiff's base. A third Schiff's base is generated by way of tautomerization. Hydrolysis of the third Schiff's base produces formaldehyde and a diketo amine. More ninhydrin reagent subsequently reacts with the amino functional group of this diketo amine to produce a new Schiff's base, which subsequently tautomerizes to produce Rhuemann's purple.

191. The product is an ester, but it is a phosphate ester instead of a carboxylate ester. The phosphorus atoms in ATP are electropositive relative to the adjacent oxygen. The terminal phosphorous atom undergoes nucleophilic attack by the $5'$-OH of the ribose to produce the desired ester. In the living cell the reaction occurs at the $5'$-OH group because of specific enzyme catalysis.

192. DNA has no oxygen in the $2'$ position of the sugar. Removal of the proton from the $2'$-OH group in RNA generates an alkoxide ion. This alkoxide is used to attack the phosphorous of the phosphate ester to produce a new cyclic $2'$, $3'$-phosphodiester and a free $5'$-OH. This cyclic ester subsequently undergoes ring opening to produce a mixture of nucleotides containing $2'$- or $3'$-phosphate esters.

193. The lowest energy MO has no nodes (i.e., the signs of the MO do not change). As the MO's increase in energy the number of nodes increase, so that the second, third, and fourth MO's have one, two, and three nodes, respectively. The energy diagram can thus be represented from low energy to highest energy, as ++++, ++¦--, +¦--¦+, and +¦-¦+¦- where the dotted lines represent the nodes. These signs represent only one side of the pi surface. The other side will have opposite sign. The frontier orbital HOMO for the ground state (thermal) reaction is (++¦--), indicating an allowed conrotation leading to cis-dimethylcyclobutene. For a photochemical reaction, the highest occupied MO is (+¦--¦+), indicating disrotation to trans-dimethylcyclobutene.

194. The only allowed thermal process that can easily occur in this system is a [2+4] cycloaddition. The butadiene-like pi system in the cyclohexadiene moiety will cycloadd to the ethylene moiety directly *above* it (see structure). This will place the deuteriums on the methylene bridge of the newly formed cyclobutane moiety in the tricyclic product.

104   VERBAL LEADS

195. The MO's from lowest energy to highest energy are (increasing nodes): +++++, ++0--, +0-0+, +-0+-, and +-+-+. The cation has four pi electrons and thus the HOMO (frontier orbital) is ++0--. This indicates that a conrotatory mode is allowed in order to bring the terminal lobes of similar sign "into contact". The anion has six pi electrons so the HOMO is +0-0+, which indicates a disrotatory mode is allowed.

196. This is simply a [2+2] cycloaddition, similar to the cycloaddition of two ethylenes to cyclobutane. In the ground state (thermal reaction) the HOMO of a one ethylene moiety is ++, and the LUMO of the other component is +-. Thus the ground state (thermal) reaction is not allowed. In the excited state, the HOMO is +-, as is the LUMO of the other (ground state) moiety (+-). Therefore this reaction is allowed. The reaction is allowed photochemically and forbidden thermally.

197. The first step is a cis pyrolytic elimination of acetic acid in which the acetoxy carbonyl oxygen atom, in a concerted process, coordinates (and then departs with) an adjacent cis hydrogen. The initial product is *trans*-3,5-dimethylcyclopentene. Bromine adds to the double bond in this cyclic alkene to give trans dibromides. The final products are dibromodimethylcyclopentanes.

198. Lithium transfers one electron to the α,β-unsaturated carbonyl system to form a radical anion. The radical anion abstracts a proton from the solvent ammonia to yield a radical. This radical reacts with lithium to give an enolate, which upon work-up gives the product.

199. Bromination with NBS should occur at the benzylic position. This compound contains a very reactive bromide (it is benzylic and secondary). The acid group adjacent to this active center can lose a proton and act as an internal nucleophile. Simultaneous loss of this acidic proton and internal nucleophilic attack on the carbon atom bearing bromine yields a five-membered lactone, the final product isolated.

200. Subtracting the partial formula of a para-disubstituted benzene fragment from the original formula ($C_9H_{10}O_2 - C_6H_4$) leaves $C_3H_6O_2$ for substituents on the ring. One of the ring substituents contains (or is) an aldehyde function (recall the + Tollens' test). This is confirmed by hemiacetal formation with CH_3OH. Thus, $C_3H_6O_2$ - CHO leaves a C_2H_5O fragment. The upfield quartet and triplet in the off-resonance decoupled spectrum are consistent with an ethoxy group (the methylene yields the triplet and the methyl yields the quartet). The original compound is thus *p*-ethoxybenzaldehyde.

201. The formula C_7H_8O shows that this molecule is highly unsaturated and/or contains rings. It is hard to conceive of a structure with enough rings to account for the lack of hydrogen, but one highly unsaturated ring (i.e., a benzene ring?) seems a good possibility. The low field 7.35 absorption in the proton spectrum (5H) is consistent with a monosubstituted benzene ring. ($C_6H_8O - C_6H_5$) (the phenyl group) leaves only CH_3O. The triplet in the ^{13}C spectrum means that there must be a $-CH_2-$ group present. Thus, $CH_3O - CH_2$ leaves only an OH fragment. The three fragments of the molecule are therefore phenyl, α-methylene, and hydroxyl. These are also consistent with the proton NMR.

202. The s character of the CH bond increases in going from ethane (sp^3) to ethene (sp^2) to ethyne (sp). Thus, ethane has a $^1J_{CH}$ of 125 Hz, ethene 156 Hz, and ethyne 248 Hz. The relationship between s character in the bond and $^1J_{CH}$ is thus approximately:

$$^1J_{CH} = 500\,X$$

where X is the fraction of s character in the bond. The $CF_3\overset{*}{C}HO$ J value of 207 is between that of ethene and ethyne [between sp^2 (0.33s) and sp(0.50s)].

203. The three ^{13}C absorptions, triplet, doublet, and singlet are indicative of secondary, tertiary, and quaternary carbon atoms, respectively, that is,

$$-CH_2- \ / -CH- \ / -\overset{|}{\underset{|}{C}}-$$

The reduction with 2 mole of hydrogen is consistent with two unsaturations (i.e., two double bonds or a triple bond). Compound C is probably an acetate, which when pyrolyzed yields propylene and acetic acid. Compound B is $CH_3CH_2CH_2OH$ and compound C is $CH_3CH_2CH_2O_2CCH_3$.

204. The reaction of compound X with sodium to yield hydrogen and an alkoxide is indicative of an alcohol. The formula of compound X is indicative of one ring or one unsaturation. The ^{13}C shift values (all at high field) and the lack of reaction with hydrogen are consistent with the presence of a ring (i.e., there are no olefinic carbon atoms). The ^{13}C–H splitting patterns show that compound X contains:

3   CH_3- groups (3 quartets)

3   $-CH_2-$ groups (3 triplets)

4   $-CH-$ groups (4 doublets)

Since Compound X is a substituted cyclohexane we can write a partial structure:

   + 3 — CH_3 groups

(Three of the cyclohexane ring carbon atoms are bonded to other carbon atoms producing the — CH — moieties)

205. The chain is symmetrical and T_1 values decrease from the terminal carbon atoms towards the center of the molecule. Motion must be easier at the end of the chain than in the middle. Motion at the ends of the molecule is easier because the carbon atoms near the end have fewer and lower weight fragments attached. Motion (rotation as well as other modes) is thus easier.

4 SOLUTIONS WITH STRUCTURAL FORMULAS

1. (a) 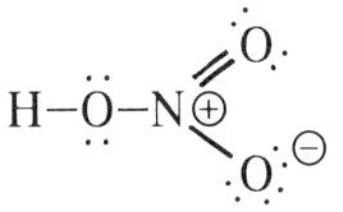 H—Ö—N⊕(=Ö:)(—Ö:⁻)

    (b) 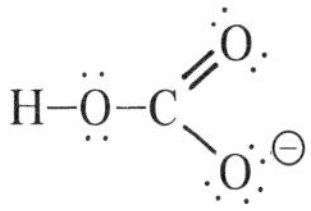 H—Ö—C(=Ö:)(—Ö:⁻)

    (c) CH₃—N⊕(=Ö:)(—Ö:⁻)

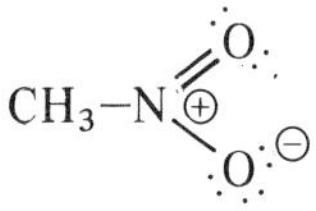

    (d) 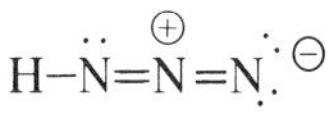 H—N̈=N⊕=N:⁻

2. Empirical formula = CH₂O
    Molecular formula = C₆H₁₂O₆

3. (a)–(d) and (f) are identical pairs, but (e) shows an ether (diethyl ether) and a ketone (acetone), respectively.

4. (a) 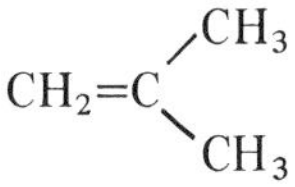 CH₂=C(CH₃)(CH₃)   H\C=C/CH₃ (with H and CH₃) → sp^3 Tetrahedral (109°)

    Planar   Trigonal planar (120°) sp^2

    (b) 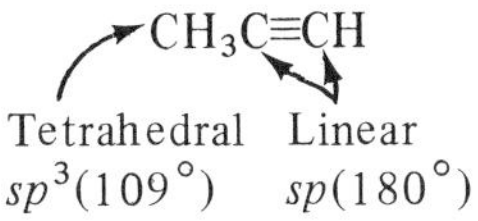 CH₃C≡CH

    Tetrahedral   Linear
    sp^3(109°) sp(180°)

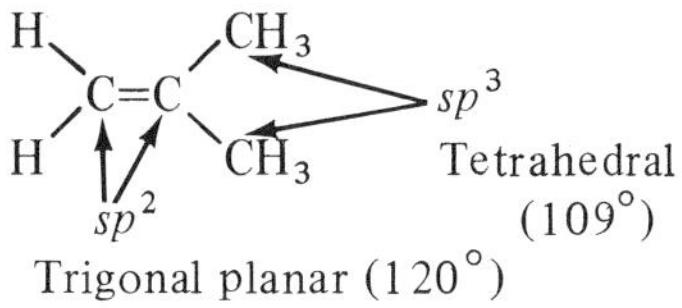

    (c) CH₃CH₂CH₃

    All bond angles are tetrahedral, sp^3 (109°).

108 SOLUTIONS WITH STRUCTURAL FORMULAS

5.

Linear
no net dipole

Trigonal planar
no net dipole

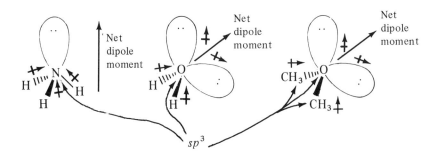

6. Structure (a) represents two resonance structures. Structure (b) represents two isomers (isomers that differ only in the placement of hydrogen atoms can also be called tautomers).

7.
Compound	Justification
(a) Butanoic acid	Hydrogen bonding
(b) 1-Butanol	Hydrogen bonding
(c) Hexylamine	Hydrogen bonding
(d) Octane	Molecular weight
(e) 1-Butanol	Hydrogen bonding
(f) *cis*-2-Butanol	Dipole moment

8. $H_2SO_4 + Et_2O \longrightarrow$ Et–$\overset{\overset{\oplus}{}}{\underset{H}{O}}$–Et + $HSO_4^{\ominus}$

This ion pair dissolves
in more sulfuric acid

9. K_a $CF_3CO_2H > CH_3CO_2H$

   pK_a $CH_3CO_2H > CF_3CO_2H$

10. $CH_3OCH_2CH_3$ $CH_3CH_2CH_2OH$ $CH_3\underset{\underset{OH}{|}}{CH}CH_3$

11. $CH_3CH_2CH_2-O-CH_2CH_2CH_3$; (iPr-O-iPr); (iPr-O-nPr);

(iBu-O-CH_3); (nBu-O-CH_3); (tBu-O-CH_3);

(iPr-CH(CH_3)-OCH_3); (iPr-CH_2-OEt); (iPr-OEt); (nPr-O-Et);

(tBu-OEt)

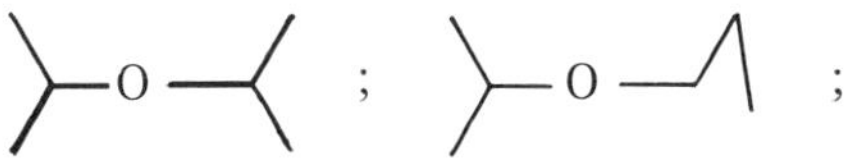

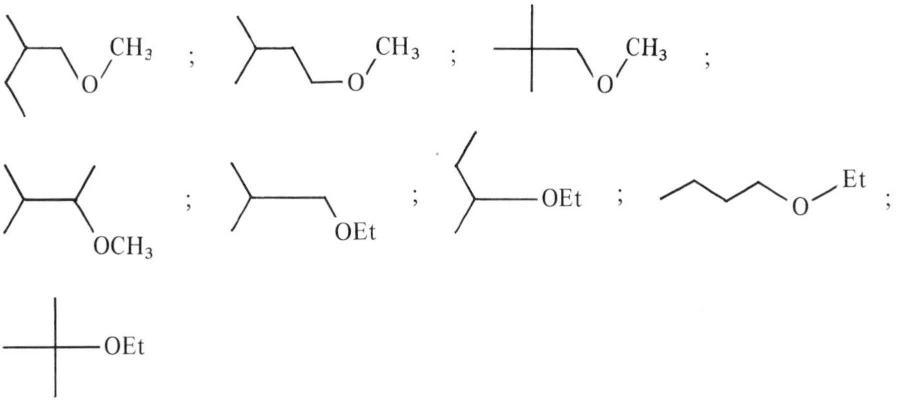

12. $CH_3-C\overset{\overset{\ominus}{O}}{\underset{O^\ominus}{\diagdown}} \longleftrightarrow CH_3-C\overset{O^\ominus}{\underset{O}{\diagdown}}$

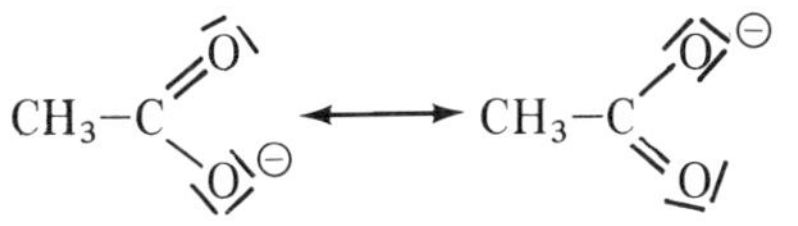

Neither of these structures is in complete accord with the physical properties of the molecule. The actual structure of the ion is best represented by the following hybrid.

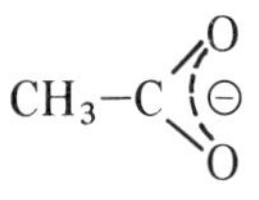

$CH_3-C\overset{O}{\underset{O}{\diagdown}}^\ominus$

13. (iPr-CH_2-Cl) ; (tBu-Cl) ; $CH_3CH_2\underset{\underset{Cl}{|}}{CH}CH_3$ $CH_3(CH_2)_3Cl$

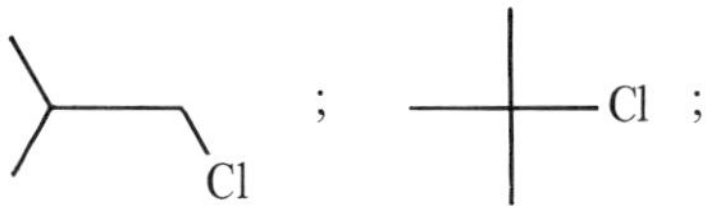

14.

Equatorial — H Cl — Axial

     H

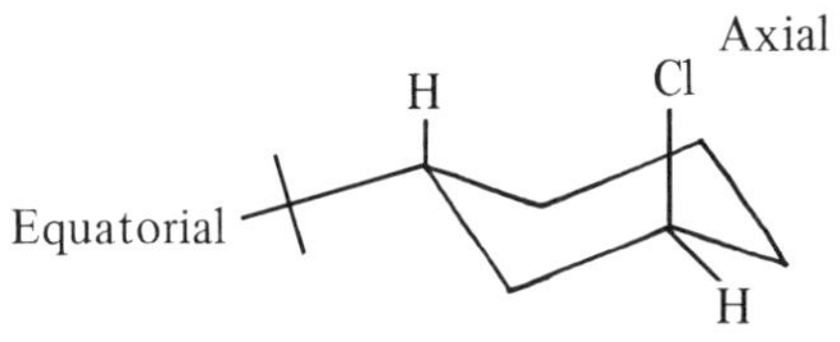

110 SOLUTIONS WITH STRUCTURAL FORMULAS

15. (a) [cyclohexane] + Br$_2$ →(Δ) [cyclohexyl-Br] →(Li) [cyclohexyl-Li] →(CuI) ([cyclohexyl]$_2$CuLi)

(b) CH$_3$CH$_3$ →(Br$_2$, Δ) CH$_3$CH$_2$Br → [reacts with (cyclohexyl)$_2$CuLi] → cyclohexyl-CH$_2$CH$_3$

Use of Et$_2$CuLi and cyclohexyl bromide in the last step does not form product because the reaction requires a primary halide.

16. (a) Bicyclo[2.1.1]hexane
(b) Bicyclo[2.2.2]octane
(c) Bicyclo[3.2.0]heptane
(d) 8-Methylbicyclo[4.3.0]nonane

17. (a) 1-Pentanol
(b) 3,3-Dimethyl-1-pentanol
(c) 4-Methyl-2-pentanol
(d) 4,4,5-Trimethyl-2-hexanol

18. (a) 2,2,4-Trimethylpentane
(b) 3-Methylhexane
(c) 3-Methyl-6-ethylnonane
(d) 3-Ethyl-3-methylhexane

19.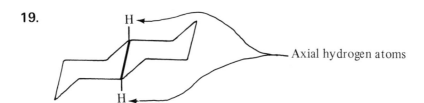
Axial hydrogen atoms

SOLUTIONS WITH STRUCTURAL FORMULAS   111

20.

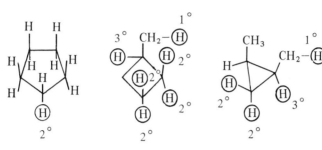

Enantiomorphic pair   Enantiomorphic pair

21.

Ⓗ = different types of replacable hydrogen atoms

22. Chain initiation: $Cl_2 \xrightarrow{h\nu} 2\;Cl\cdot$

Chain propagation: $Cl\cdot + CH_3CH_3 \longrightarrow CH_3CH_2\cdot + HCl$

$CH_3CH_2\cdot + Cl_2 \longrightarrow CH_3CH_2Cl + Cl\cdot$

Chain termination: $2\;Cl\cdot \longrightarrow Cl_2$

$CH_3CH_2\cdot + Cl\cdot \longrightarrow CH_3CH_2Cl$

23.

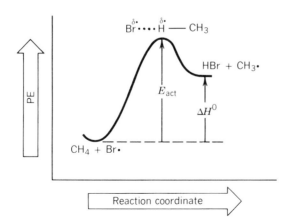

112 SOLUTIONS WITH STRUCTURAL FORMULAS

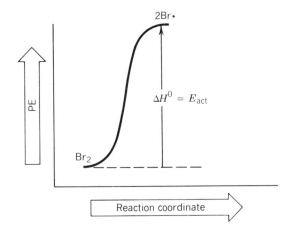

24. The production of isopropyl bromide predominates over propyl bromide because it has a more facile rate-determining propagation step.

25.

$$CH_3CHCH_2CH_3 \xrightarrow[300°C]{Cl_2} \underset{Cl}{CH_2CHCH_2CH_3} + \underset{Cl}{CH_3CCH_2CH_3} + \underset{Cl}{CH_3CHCHCH_3}$$
(with CH_3 on the central/branch carbon)

$$+ \underset{Cl}{CH_3CHCH_2CH_2}$$

$$\downarrow Br_2/h\nu, \ 127°C$$

$$\underset{Br}{CH_3CCH_2CH_3} + \underset{Br}{CH_3CHCHCH_3}$$

Bromine is more selective because the Br· radical is less reactive. It seeks out bonds most easily broken.

26. Ph–C(CH$_3$)$_2$–Br + HBr

27. $CH_3O^\ominus > {^\ominus}OH > CH_3C(=O)O^\ominus > CH_3OH > H_2O$

28.

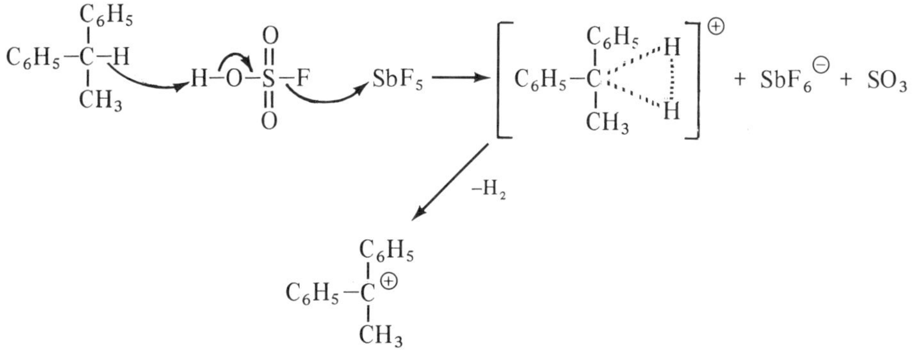

$$C_6H_5-\underset{\underset{CH_3}{|}}{\overset{\overset{C_6H_5}{|}}{C}}-H \quad H-O-\underset{\underset{O}{||}}{\overset{\overset{O}{||}}{S}}-F \xrightarrow{SbF_5} \left[C_6H_5-\underset{\underset{CH_3}{|}}{\overset{\overset{C_6H_5}{|}}{C}}\cdots\overset{H}{\underset{H}{\cdots}} \right]^{\oplus} + SbF_6^{\ominus} + SO_3$$

$$\xrightarrow{-H_2} C_6H_5-\underset{\underset{CH_3}{|}}{\overset{\overset{C_6H_5}{|}}{C}}^{\oplus}$$

29. $R-Br \xrightarrow{slow} R^{\oplus} + Br^{\ominus}$

$R^{\oplus} + {}^{\ominus}OH \xrightarrow{fast} ROH$

Net reaction $R-Br + {}^{\ominus}OH \longrightarrow ROH + Br^{\ominus}$

30. $(CH_3)_2CH-\underset{\underset{Br}{|}}{C}(CH_3)_2 \xrightarrow{HCO_2H} (CH_3)_2CH-\overset{\overset{CH_3}{|}}{\underset{\underset{CH_3}{|}}{C^{\oplus}}} + Br^{\ominus}$

elimination ↙   substitution ↓   ↖ $H-C\overset{O-H}{\underset{O}{\diagdown}}$

$(CH_3)_2C=C(CH_3)_2$

$(CH_3)_2CH-C(CH_3)_2$

$HBr + (CH_3)_2CH-\underset{\underset{O}{|}}{C}(CH_3)_2 \longleftarrow Br^{\ominus} \curvearrowright H-\overset{\oplus}{O}$
$ \underset{H}{\overset{|}{C}}=O \underset{H}{\overset{|}{C}}\diagdown_O$

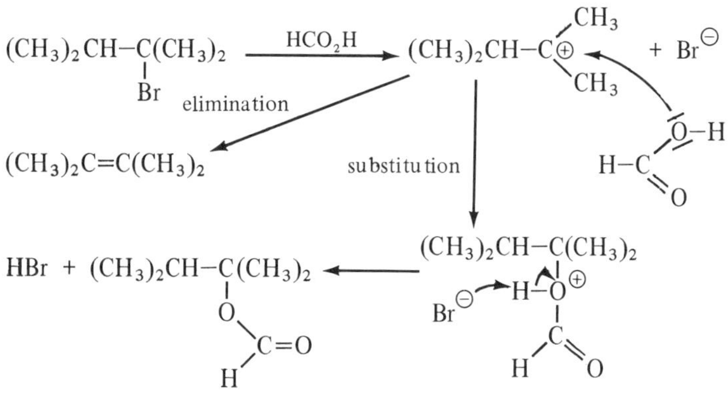

31. [Mechanism showing CH_3-S-H attacking a tosylate with displacement of $CH_3-C_6H_4-SO_3^{\ominus}$, giving $CH_3-\overset{\oplus}{\underset{H}{S}}-C(CH_3)(H)(C_6H_5)$ → product]

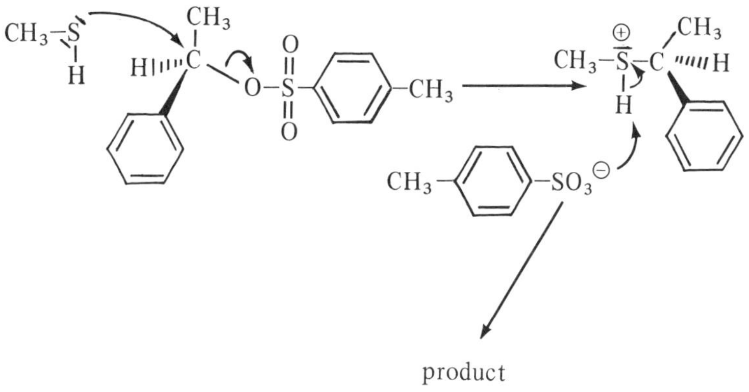

32.

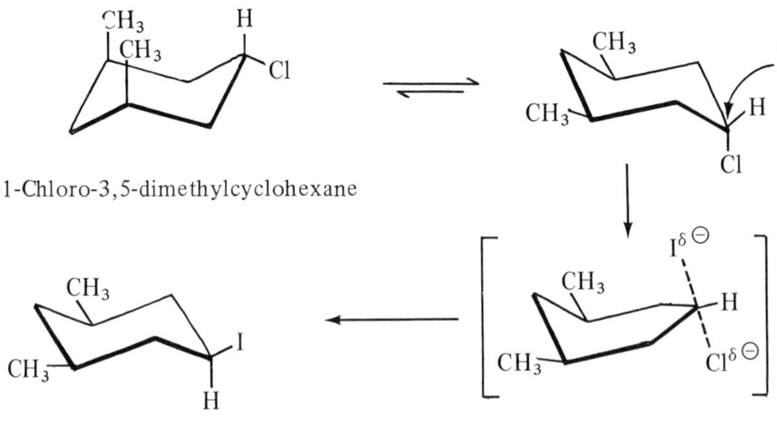

trans-1-Chloro-3,5-dimethylcyclohexane

cis-1-Iodo-3,5-dimethylcyclohexane

33. $CH_3CH_2Br: + Ag^{\oplus} \rightleftharpoons CH_3CH_2-\overset{\oplus}{Br}-Ag \quad H\underline{O}CH_2CH_3$

$$\text{diethyl ether} \xleftarrow{-H^{\oplus}} CH_3CH_2-\underset{H}{\overset{\oplus}{O}}-CH_2CH_3 + AgBr \quad (1)$$

$$CH_3CH_2OH + HBr \rightleftharpoons CH_3CH_2-\overset{\oplus}{O}H_2 \longrightarrow CH_3CH_2Br + H_2O \quad (2)$$
$$Br^{\ominus}$$

34.

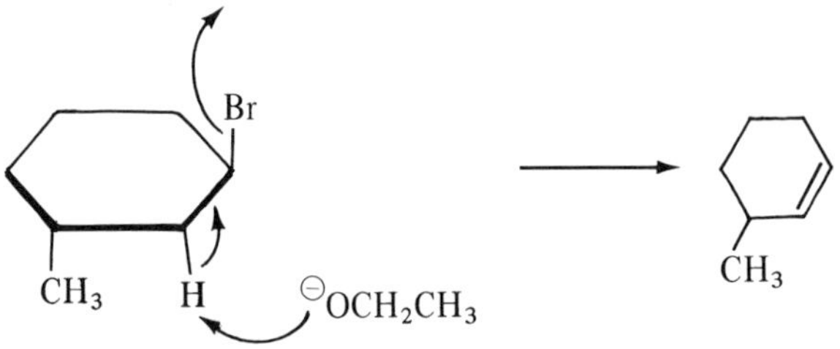

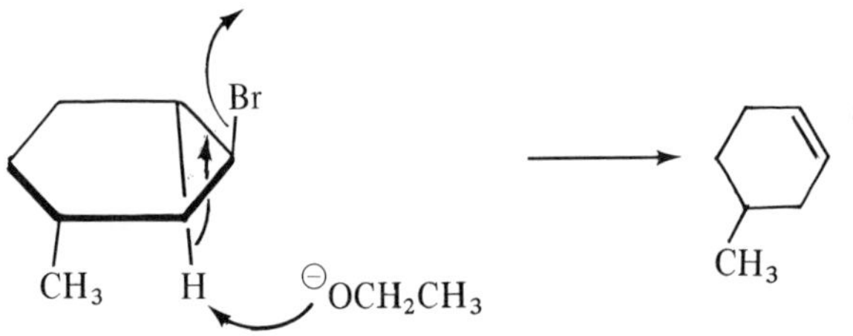

SOLUTIONS WITH STRUCTURAL FORMULAS 115

35. $(CH_3)_3C-Cl \xrightarrow{slow} (CH_3)_3C^{\oplus} + Cl^{\ominus}$

$$\underset{H}{\overset{H_2\ddot{O}:\text{ or } CH_3\ddot{O}:}{\underset{|}{H}}} \quad CH_2-\overset{CH_3}{\underset{CH_3}{\overset{|}{C^{\oplus}}}} \xrightarrow{fast} CH_2=\overset{CH_3}{\underset{CH_3}{C}} + H_3O^{\oplus} \text{ or } CH_3\overset{\oplus}{O}H_2$$

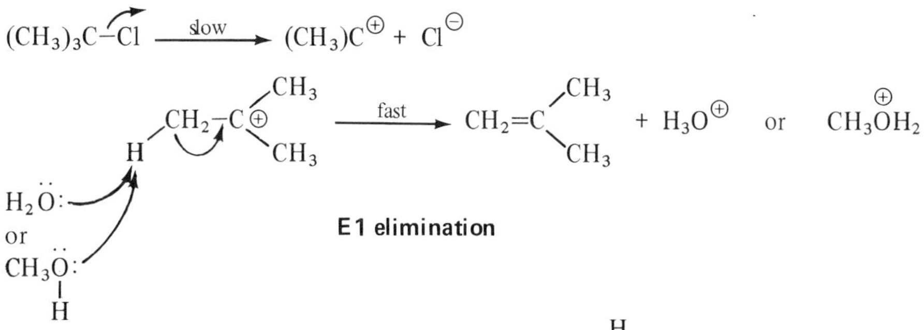

E1 elimination

$$(CH_3)_3C^{\oplus} + :OH_2 \xrightarrow{fast} (CH_3)_3C-\overset{H}{\underset{H}{\overset{\oplus}{O}}} \xleftarrow{H_2O:}$$

$$\xrightarrow{fast} H_3O^{\oplus} + (CH_3)_3OH$$

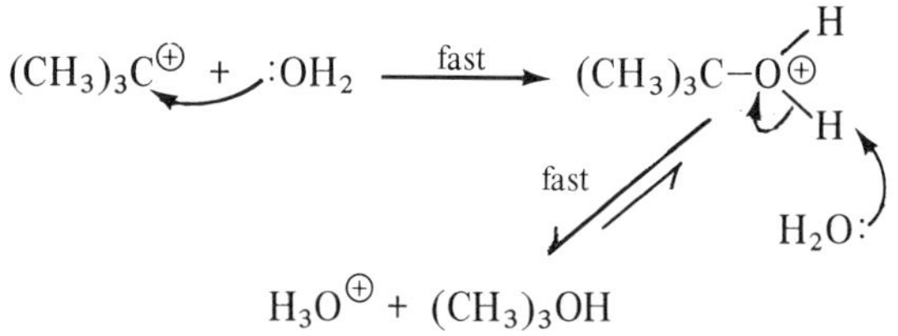

S$_N$1 substitution

A similar S$_N$1 substitution on the carbocation by CH$_3$OH yields:

$(CH_3)_3-O-CH_3$

36. $(CH_3)_2CH-O^{\ominus} \quad H-CH_2-\underset{C(CH_3)_3}{\overset{CH_3}{\overset{|}{C}}}-Br \longrightarrow (CH_3)_2CHOH + CH_2=\underset{C(CH_3)_3}{\overset{CH_3}{C}} + Br^{\ominus}$

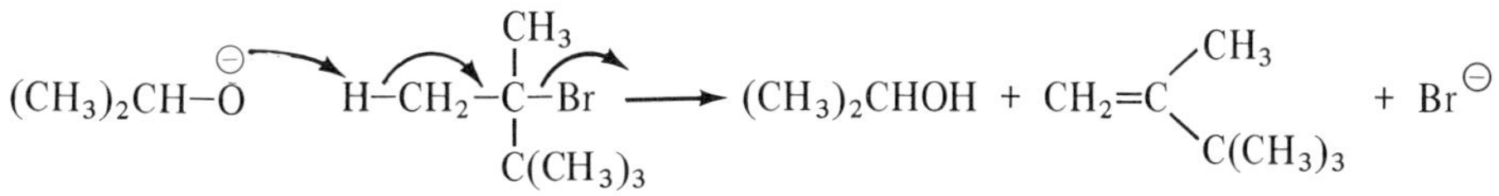

37.

[Reaction showing benzyl-N(CH$_3$)(CH$_2$CH$_3$) attacking C-S$^{\oplus}$ with phenyl groups, Cl$^{\ominus}$, yielding quaternary ammonium product with CD and diphenyl sulfide]

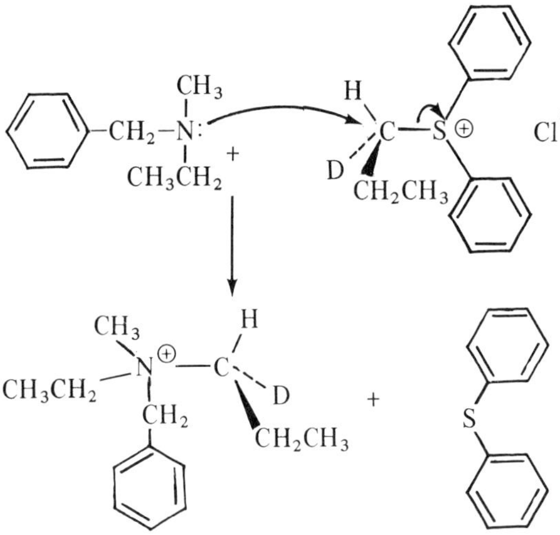

38.

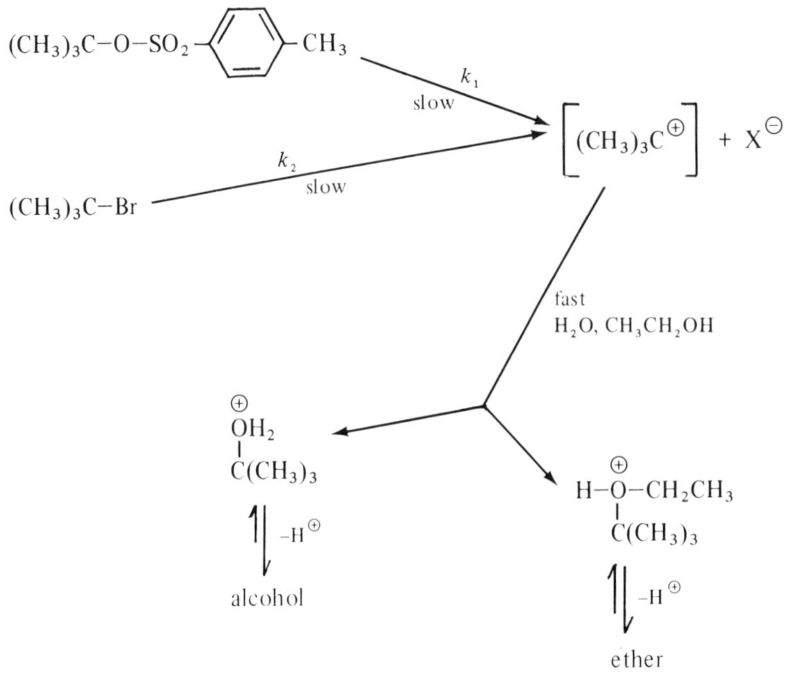

$(CH_3)_3C-O-SO_2-C_6H_4-CH_3$
k_1
slow
$(CH_3)_3C-Br$
k_2
slow
$[(CH_3)_3C^{\oplus}] + X^{\ominus}$
fast
H_2O, CH_3CH_2OH
$\overset{\oplus}{O}H_2$
$C(CH_3)_3$
$-H^{\oplus}$
alcohol
$H-\overset{\oplus}{O}-CH_2CH_3$
$C(CH_3)_3$
$-H^{\oplus}$
ether

39.

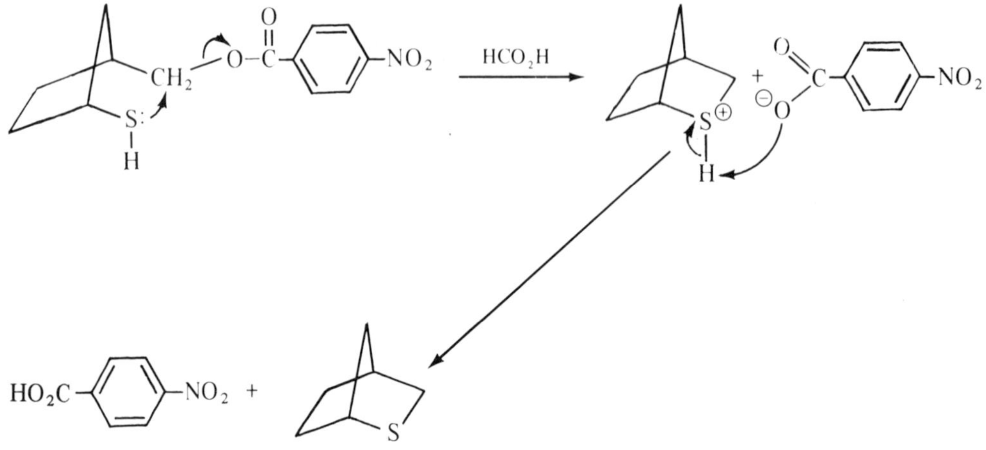

O
CH_2
O–C
NO_2
S:
H
HCO_2H
+
S⊕
H
O
C
NO_2
⊖O
HO_2C
NO_2
+
S

40. Ion B is most stable due to the resonance interaction shown in the following.

$$\left[\begin{array}{c} CH_3 \\ \\ CH_3 \end{array} \!\!\!N\!\!-\!\!\!\bigcirc\!\!\!-\overset{\oplus}{C}H_2 \longleftrightarrow \begin{array}{c} CH_3 \\ \\ CH_3 \end{array}\!\!\!\overset{\oplus}{N}\!\!=\!\!\bigcirc\!\!=\!\!CH_2 \right]$$

41

$$(CH_3)_2\overset{\curvearrowright}{C}\!\!-\!\!Br \xrightarrow[\text{① major}]{CH_3O^{\ominus}} \begin{array}{c} CH_3 \diagdown \diagup CH_3 \\ C \\ \parallel \\ C \\ H \diagup \diagdown CH_3 \end{array} \quad (1)$$

$$CH_3O^{\ominus} \curvearrowright H\!\!-\!\!\overset{|}{\underset{CH_3}{C}H}$$

Most stable

$$\underset{CH_3O^{\ominus}}{H\!\!-\!\!CH_2\!\!-\!\!\overset{\overset{CH_3}{|}}{\underset{\underset{CH_3}{|}}{C}}\!\!-\!\!Br} \xrightarrow[\text{① minor}]{CH_3O^{\ominus}} H_2C\!\!=\!\!C\!\!\begin{array}{c} \diagup CH_3 \\ \diagdown CH_2CH_3 \end{array} \quad (2)$$

Least stable

$$CH_3\!\!-\!\!\overset{\overset{CH_3}{|}}{\underset{\underset{CH_3}{|}}{C}}\!\!-\!\!O^{\ominus} \quad CH_3\!\!-\!\!\overset{\overset{CH_3}{|}}{\underset{\underset{CH_3}{|}}{C}}\!\!-\!\!Br \xrightarrow[\text{② minor}]{t\text{-BuO}^{\ominus}} \begin{array}{c} CH_3 \diagdown \diagup CH_3 \\ C \\ \parallel \\ C \\ H \diagup \diagdown CH_3 \end{array}$$

Most hindered

$$CH_3\!\!-\!\!\overset{\overset{CH_3}{|}}{\underset{\underset{CH_3}{|}}{C}}\!\!-\!\!O^{\ominus} \quad H\!\!-\!\!CH_2\!\!-\!\!\overset{\overset{CH_3}{|}}{\underset{\underset{\underset{CH_3}{|}}{CH_2}}{C}}\!\!-\!\!Br \xrightarrow[\text{② major}]{t\text{-BuO}^{\ominus}} H_2C\!\!=\!\!C\!\!\begin{array}{c} \diagup CH_3 \\ \diagdown CH_2CH_3 \end{array}$$

Least hindered

118 SOLUTIONS WITH STRUCTURAL FORMULAS

42. $(CH_3)_3C-CH_2-\ddot{O}H \quad H^\oplus \longrightarrow CH_3-\underset{\underset{CH_3}{|}}{\overset{\overset{CH_3}{|}}{C}}-CH_2-\overset{\oplus}{O}H_2$

$\downarrow -H_2O$

$\underset{H_a}{\overset{CH_3}{\underset{|}{\overset{|}{CH_2-\overset{\oplus}{C}-CH-CH_3}}}}$

$\overset{-H_a^\oplus}{\longleftarrow} \quad \underset{H_b}{}$

$\underset{CH_3}{\overset{CH_3}{C=C}}\underset{CH_3}{\overset{H}{}}$

Major

$\overset{-H_b^\oplus}{\swarrow} \qquad H_2O:$

$\underset{H_2C}{\overset{CH_3}{C-CH_2CH_3}}$

Minor

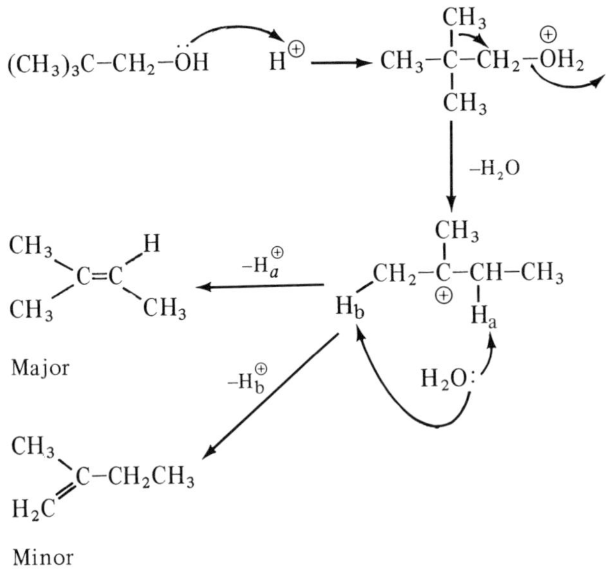

43. Some of the possibilities are

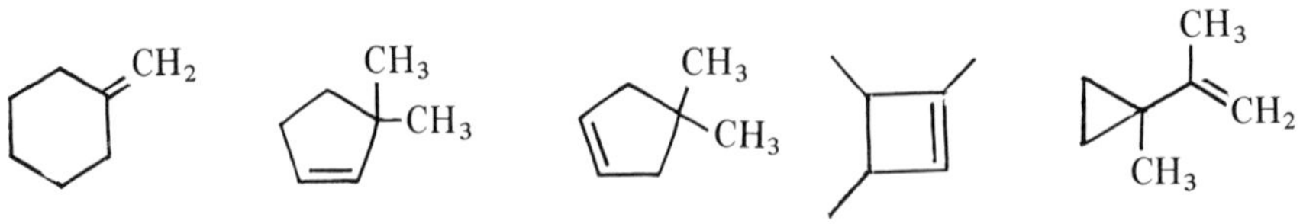

[structures: methylenecyclohexane; 1,5-dimethylcyclopentadiene variant; 1,1-dimethylcyclopentene variant; dimethylcyclobutene; isopropenylcyclopropane variant]

44.

[diagrams: tip → rotate → finish, showing stereochemical manipulation of C with CH₃, H, Br, Cl substituents]

tip $\longrightarrow$ rotate $\longrightarrow$ finish

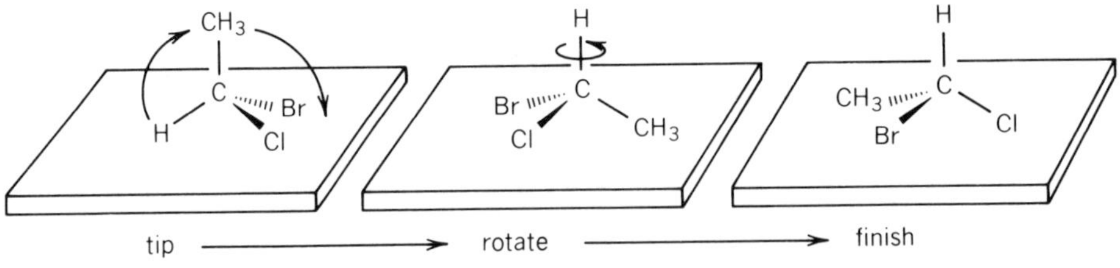

45.

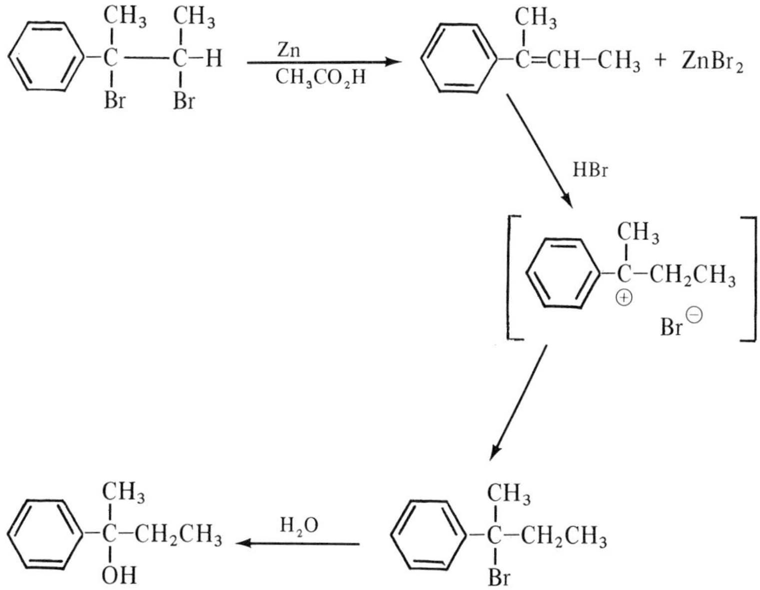

$$\underset{\underset{\text{Br}}{|}}{\overset{\overset{\text{CH}_3}{|}}{\text{Ph-C}}}-\underset{\underset{\text{Br}}{|}}{\overset{\overset{\text{CH}_3}{|}}{\text{C}}}-\text{H} \xrightarrow[\text{CH}_3\text{CO}_2\text{H}]{\text{Zn}} \text{Ph-}\overset{\overset{\text{CH}_3}{|}}{\text{C}}=\text{CH-CH}_3 + \text{ZnBr}_2$$

$$\downarrow \text{HBr}$$

$$\left[\text{Ph-}\overset{\overset{\text{CH}_3}{|}}{\underset{\oplus}{\text{C}}}-\text{CH}_2\text{CH}_3 \quad \text{Br}^{\ominus} \right]$$

$$\text{Ph-}\underset{\underset{\text{OH}}{|}}{\overset{\overset{\text{CH}_3}{|}}{\text{C}}}-\text{CH}_2\text{CH}_3 \xleftarrow{\text{H}_2\text{O}} \text{Ph-}\underset{\underset{\text{Br}}{|}}{\overset{\overset{\text{CH}_3}{|}}{\text{C}}}-\text{CH}_2\text{CH}_3$$

46. (a) $CH_3CH_2CH_2CHClCH_3$

(b) $CH_3CH_2CH_2\underset{\underset{OSO_3H}{|}}{CH}CH_3$

(c) $CH_3(CH_2)_4Br$

(d) $CH_3CH_2CH_2\overset{\overset{H}{|}}{C}=O$

 $+$

 $\underset{H}{\overset{H}{\diagdown}}C=O$

(e) 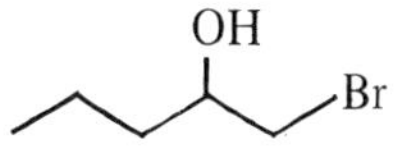 CH$_3$CH$_2$CH$_2$CH(OH)CH$_2$Br

47.

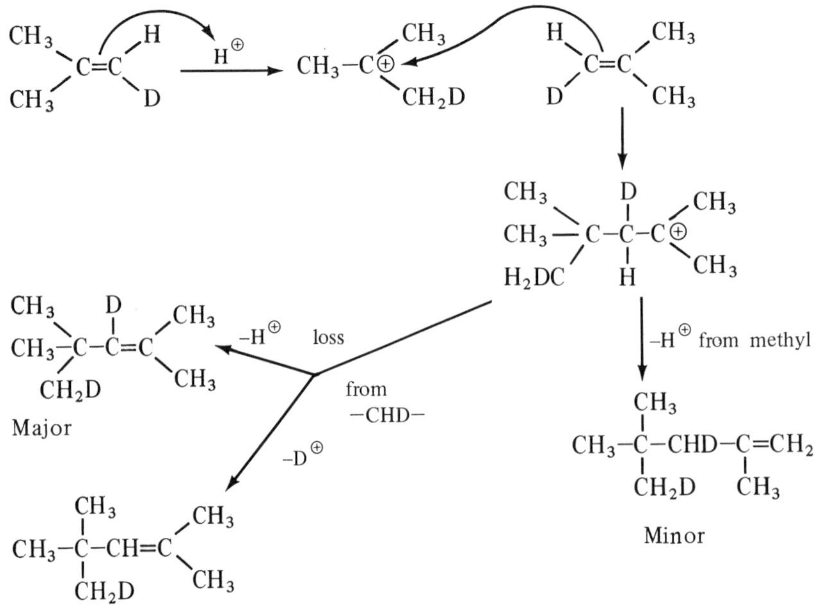

$$CH_3\text{-}C(CH_3)=CHD \xrightarrow{H^\oplus} CH_3\text{-}C^\oplus(CH_3)\text{-}CH_2D \xleftarrow{} \underset{D}{\overset{H}{C}}=\underset{CH_3}{\overset{CH_3}{C}}$$

gives cation: $CH_3\text{-}C(CH_3)(CH_2DC)\text{-}C(D)(H)\text{-}C^\oplus(CH_3)_2$

- $-H^\oplus$ loss from $-CHD-$ → Major: $CH_3\text{-}C(CH_3)(CH_2D)\text{-}C(D)=C(CH_3)_2$
- $-D^\oplus$ → $CH_3\text{-}C(CH_3)(CH_2D)\text{-}CH=C(CH_3)_2$
- $-H^\oplus$ from methyl → Minor: $CH_3\text{-}C(CH_3)(CH_2D)\text{-}CHD\text{-}C(CH_3)=CH_2$

48.

$CH_3\text{-}C(=O)\text{-}O\text{-}Hg\text{-}O\text{-}C(=O)\text{-}CH_3$ + cyclohexene →

[cyclohexyl-Hg-O-C(=O)-CH$_3$ with $\oplus$ on ring, attacked by $CH_3\text{-}\overset{..}{O}\text{-}H$] OAc$^\ominus$

→ cyclohexane with Hg-O-C(=O)-CH$_3$ and $\overset{\oplus}{O}(H)\text{-}CH_3$ substituents

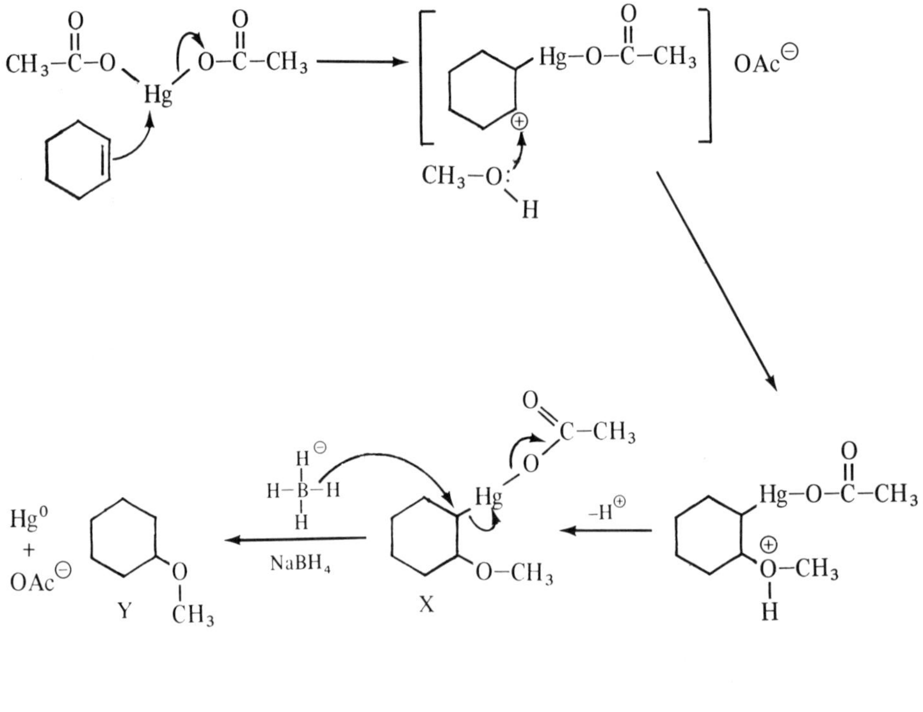

$\xrightarrow{-H^\oplus}$ **X**: cyclohexane with Hg-O-C(=O)-CH$_3$ (acetate attacking Hg) and O-CH$_3$

$\xrightarrow[NaBH_4]{H\text{-}\overset{\ominus}{B}(H)\text{-}H}$ **Y**: cyclohexane with O-CH$_3$ + Hg^0 + OAc$^\ominus$

49. $(CH_3)_2C=CH_2$ $\xrightarrow[\substack{\text{single step} \\ \text{four-center} \\ \text{transition state;} \\ \text{boron goes to least} \\ \text{hindered carbon}}]{}$ $(CH_3)_2CH-CH_2-BH_2$

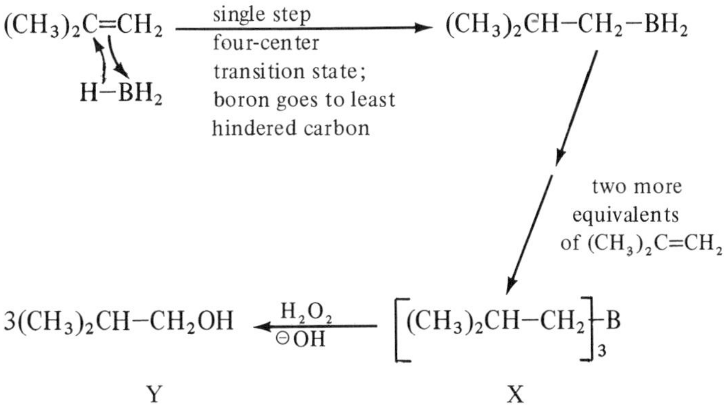

with $H-BH_2$

$\xrightarrow[\text{of }(CH_3)_2C=CH_2]{\text{two more equivalents}}$

$3(CH_3)_2CH-CH_2OH$ $\xleftarrow[\ominus OH]{H_2O_2}$ $[(CH_3)_2CH-CH_2]_3B$

Y   X

50. 3 cyclohexylidene=CH₂ (X) $\xrightarrow{BH_3}$ [cyclohexyl–CH₂]₃B

X $\xrightarrow{Pt/D_2}$ cyclohexyl-CH₂D with D

X $\xrightarrow{Pt/H_2}$ cyclohexyl-CH₃ with H

[cyclohexyl–CH₂]₃B $\xrightarrow{\Delta, CH_3CO_2D}$ cyclohexyl-CH₂D with H

51. (a) (ethylidene/isopropylidene cyclopentane structure)

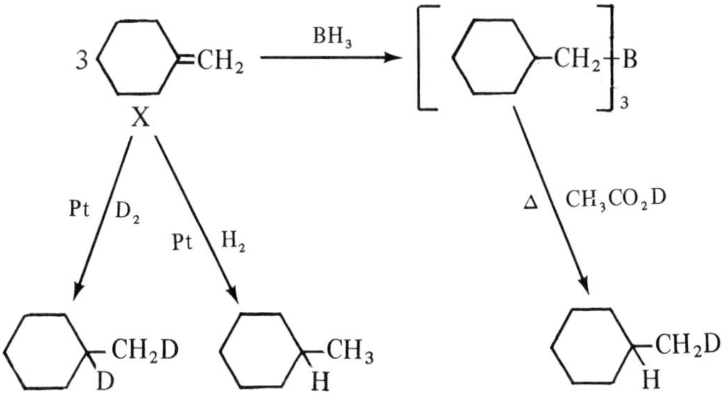

(b) (octahydronaphthalene with central double bond)

(c) (bicyclopentylidene)

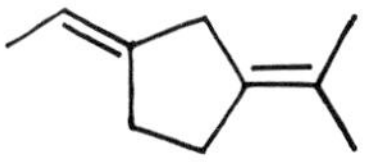

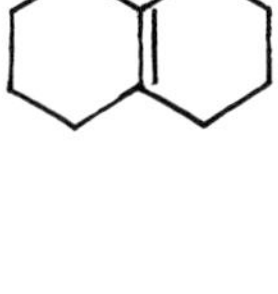

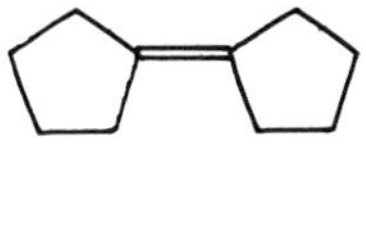

122 SOLUTIONS WITH STRUCTURAL FORMULAS

52.

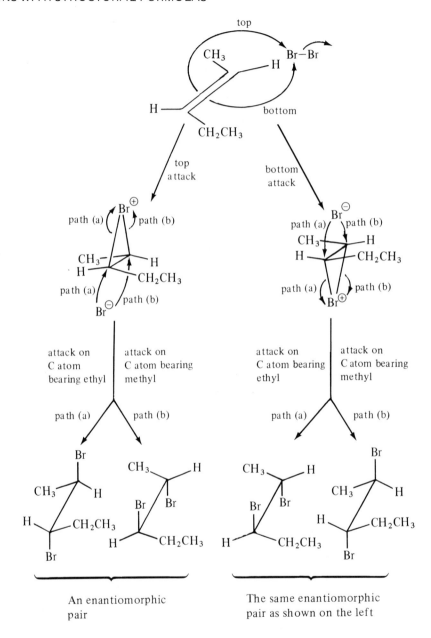

53.

54.

124 SOLUTIONS WITH STRUCTURAL FORMULAS

55.

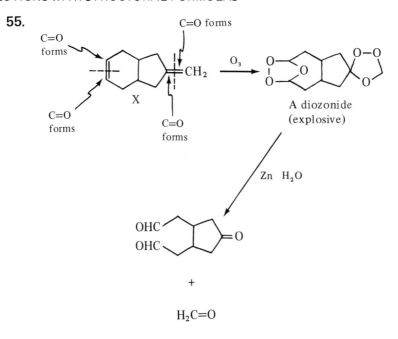

56.

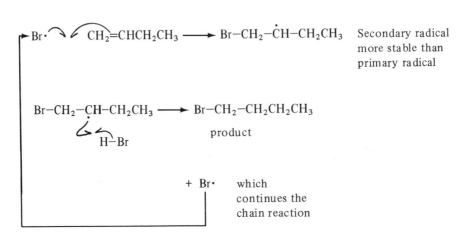

57.

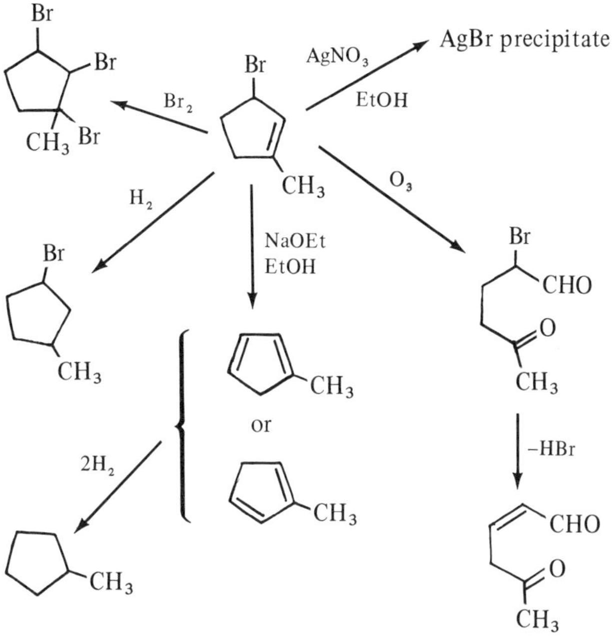

58. RO• + HBr ⟶ Br• + ROH

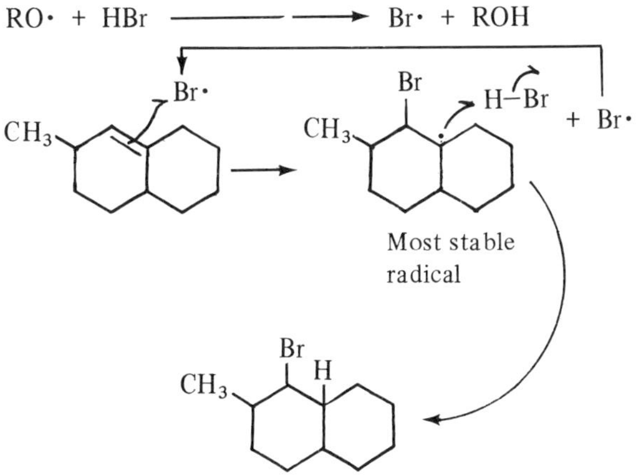

(no stereochemistry is implied, i.e., with respect to the ring fusion)

Most stable radical

126 SOLUTIONS WITH STRUCTURAL FORMULAS

59.

An enantiomorphic pair

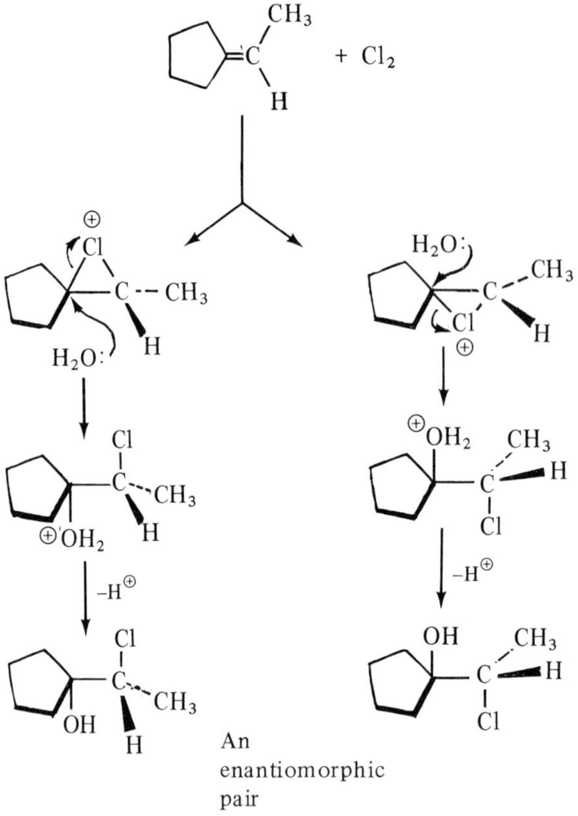

60.

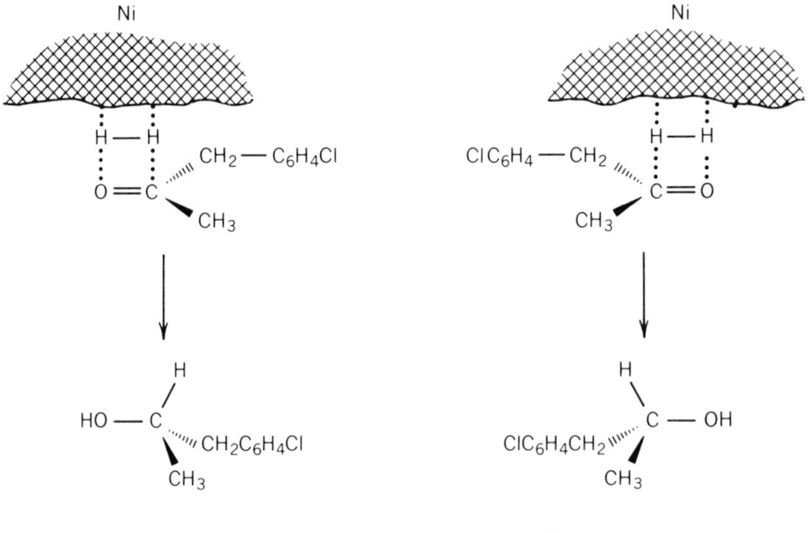

(R) Configuration   (S) Configuration

61.

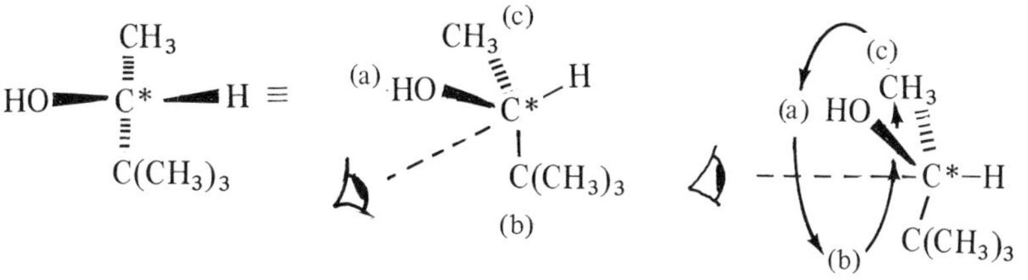

$$\text{HO}-\overset{\text{CH}_3}{\underset{\text{C(CH}_3)_3}{\text{C}^*}}-\text{H} \equiv$$

(a) HO, (c) CH$_3$, (b) C(CH$_3)_3$, H on C*

Arrows are clockwise therefore the enantiomer is (R)

62.

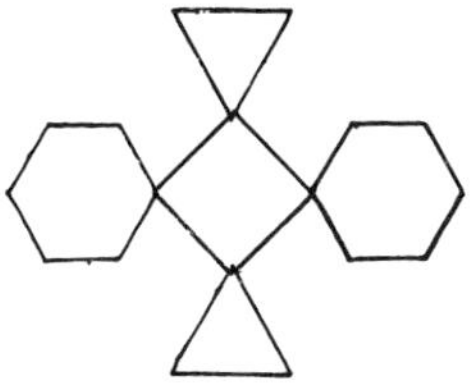

[structure: two cyclohexane rings and two cyclopropane rings spiro-fused through a central carbon]

63.

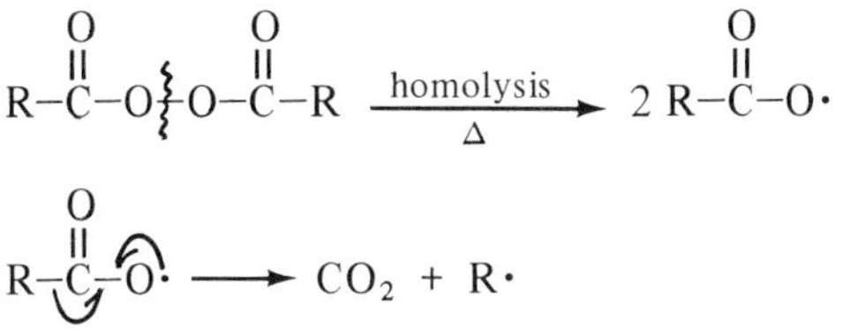

$$R-\overset{O}{\overset{\|}{C}}-O\{O-\overset{O}{\overset{\|}{C}}-R \xrightarrow[\Delta]{\text{homolysis}} 2\ R-\overset{O}{\overset{\|}{C}}-O\cdot$$

$$R-\overset{O}{\overset{\|}{C}}-O\cdot \longrightarrow CO_2 + R\cdot$$

Initiation

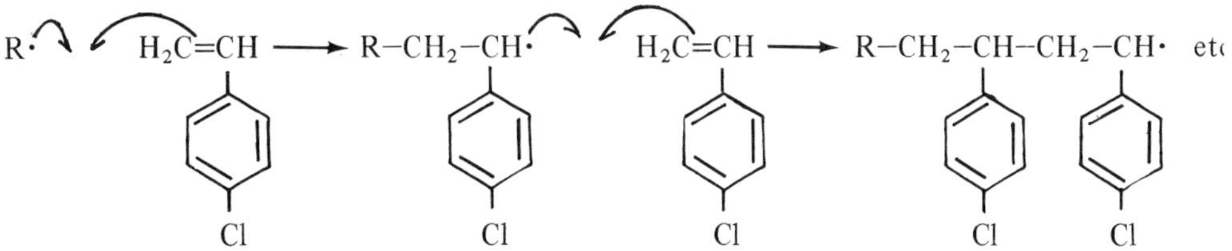

$$R\cdot + H_2C=CH(C_6H_4Cl) \longrightarrow R-CH_2-CH\cdot(C_6H_4Cl) \xrightarrow{H_2C=CH(C_6H_4Cl)} R-CH_2-CH(C_6H_4Cl)-CH_2-CH\cdot(C_6H_4Cl)\ \text{etc}$$

Chain Propagation

$$R\text{-}[CH_2\text{-}CH(C_6H_4Cl)]_n\text{-}CH_2\text{-}CH\cdot(C_6H_4Cl) + R'\cdot \longrightarrow R\text{-}[CH_2\text{-}CH(C_6H_4Cl)]_n\text{-}CH_2\text{-}CH(C_6H_4Cl)\text{-}R'$$

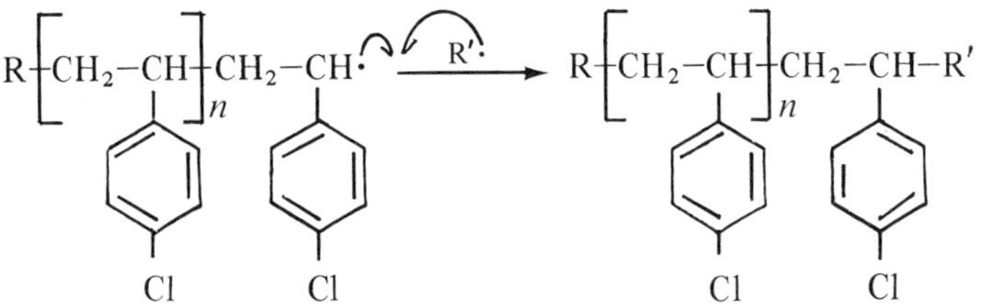

Chain Termination

128 SOLUTIONS WITH STRUCTURAL FORMULAS

64. Anionic Polymerization

$$Nu:^{\ominus} \longrightarrow \underset{R}{\overset{R}{>}}C=C\underset{R'}{\overset{R'}{<}} \longrightarrow Nu-\underset{R}{\overset{R}{|}}C-\underset{R'}{\overset{R'}{|}}C^{\ominus} \longrightarrow \underset{R}{\overset{R}{>}}C=C\underset{R'}{\overset{R'}{<}} \longrightarrow \text{etc., to polymer}$$

Cationic Polymerization

$$E^{\oplus} \longrightarrow \underset{R}{\overset{R}{>}}C=C\underset{R'}{\overset{R'}{<}} \longrightarrow E-\underset{R}{\overset{R}{|}}C-\underset{R'}{\overset{R'}{|}}C^{\oplus} \longleftarrow \underset{R}{\overset{R}{>}}C=C\underset{R'}{\overset{R'}{<}} \longrightarrow \text{etc., to polymer}$$

Radical Polymerization

$$R''\cdot \longrightarrow \underset{R}{\overset{R}{>}}\overset{..}{C}-\overset{.}{C}\underset{R'}{\overset{R'}{<}} \longrightarrow R''-\underset{R}{\overset{R}{|}}C-\underset{R'}{\overset{R'}{|}}\overset{.}{C} \longrightarrow \underset{R}{\overset{R}{>}}C=C\underset{R'}{\overset{R'}{<}} \longrightarrow \text{etc., to polymer}$$

65. $:CH_2-\overset{\oplus}{N}\equiv N: \xrightarrow{h\nu} \uparrow\downarrow CH_2 + N_2$

Singlet

$$\underset{CH_3}{\overset{HC=CH}{>}}CH_2CH_3 \xrightarrow[\text{singlet carbene insertion}]{\text{liq. phase}} \triangle \; CH_3 \; CH_2CH_3$$

$$\underset{CH_3}{\overset{HC=CH}{>}}CH_2CH_3 \xrightarrow[\text{in gas phase followed by insertion}]{\uparrow\downarrow CH_2 \longrightarrow \uparrow\uparrow CH_2} \cdots \rightleftharpoons \cdots$$

$$\xrightarrow{\text{closure}} \triangle \; CH_2CH_3 \;\; + \;\; \triangle \; CH_3 \; CH_2CH_3$$
$$CH_3$$

66.
$$\underset{\underset{\text{6 Parts}}{}}{CH_3CH_2\overset{\overset{CH_3}{|}}{CH}-CH_2CH_3} \quad \underset{\underset{\sim 1\text{ Part}}{}}{CH_3-\overset{\overset{CH_3}{|}}{\underset{\underset{CH_3}{|}}{C}}-CH_2CH_3} \quad \underset{\underset{2\text{ Parts}}{}}{CH_3\overset{\overset{CH_3}{|}}{CH}-CH(CH_3)_2}$$

$$(CH_3)_2CH-CH_2CH_2CH_3$$

3 Parts

SOLUTIONS WITH STRUCTURAL FORMULAS

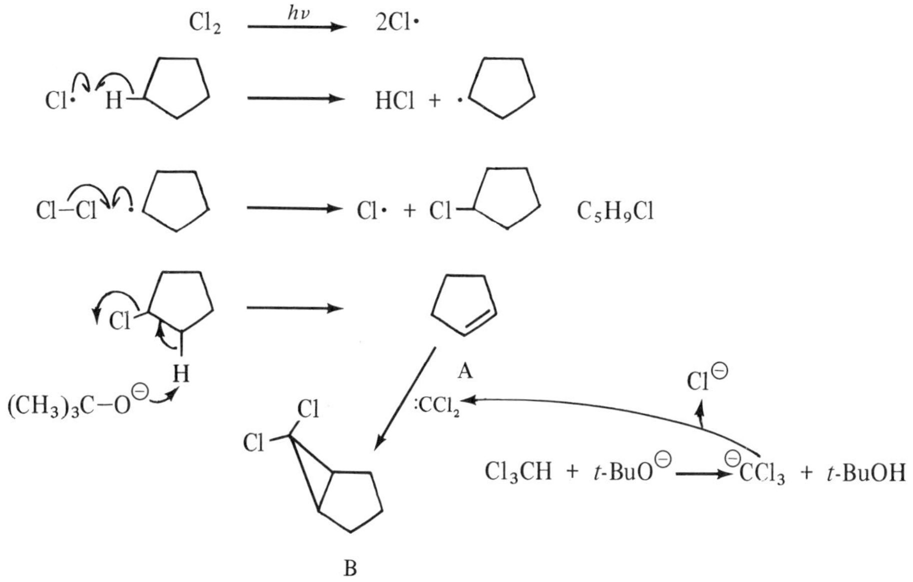

67.

$$Cl_2 \xrightarrow{h\nu} 2Cl\cdot$$

Cl· + cyclopentane-H → HCl + cyclopentyl·

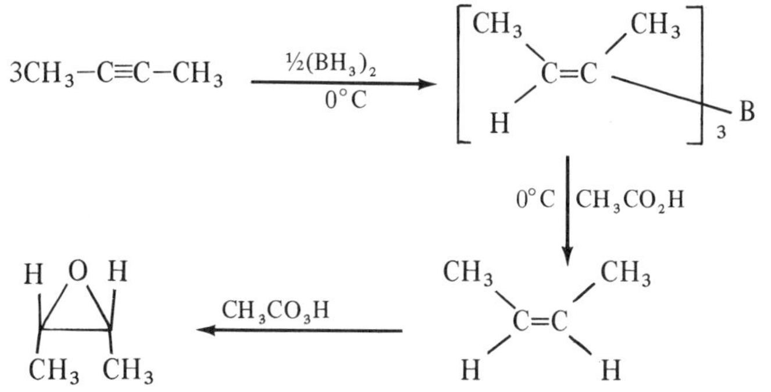

Cl–Cl + cyclopentyl· → Cl· + Cl-cyclopentane (C_5H_9Cl)

$(CH_3)_3C-O^{\ominus}$ + chlorocyclopentane → cyclopentene (A)

A + :CCl_2 → bicyclic dichloride (B)

$Cl_3CH + t\text{-}BuO^{\ominus} \longrightarrow {}^{\ominus}CCl_3 + t\text{-}BuOH$

${}^{\ominus}CCl_3 \longrightarrow :CCl_2 + Cl^{\ominus}$

68.

$$3CH_3-C{\equiv}C-CH_3 \xrightarrow[0°C]{½(BH_3)_2} \left[\begin{array}{c} CH_3 \quad\quad CH_3 \\ \diagdown\;\;\;\;\diagup \\ C{=}C \\ \diagup\;\;\;\;\diagdown \\ H \quad\quad\quad\quad \end{array} \right]_3 B$$

$\xrightarrow[0°C]{CH_3CO_2H}$ cis-2-butene (CH_3, CH_3 cis; H, H cis)

$\xrightarrow{CH_3CO_3H}$ cis Product (epoxide with H, CH_3, CH_3, H)

$CH_3-C{\equiv}C-CH_3 \xrightarrow[C_2H_5NH_2 \;\; -78°C]{Li}$ trans-2-butene

$\xrightarrow{CH_3CO_3H}$ trans Product (epoxide with H, CH_3 on one side; CH_3, H on other)

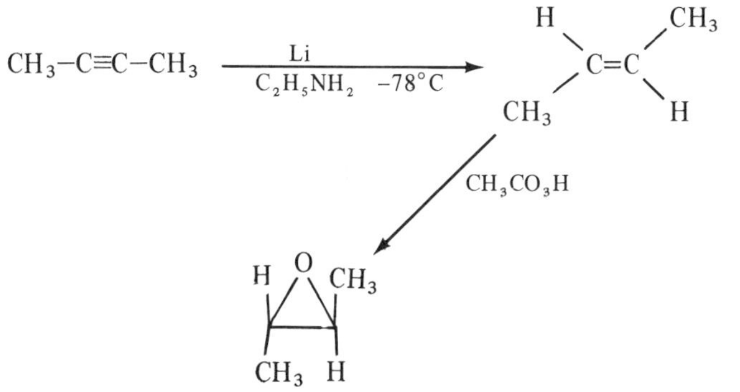

69.

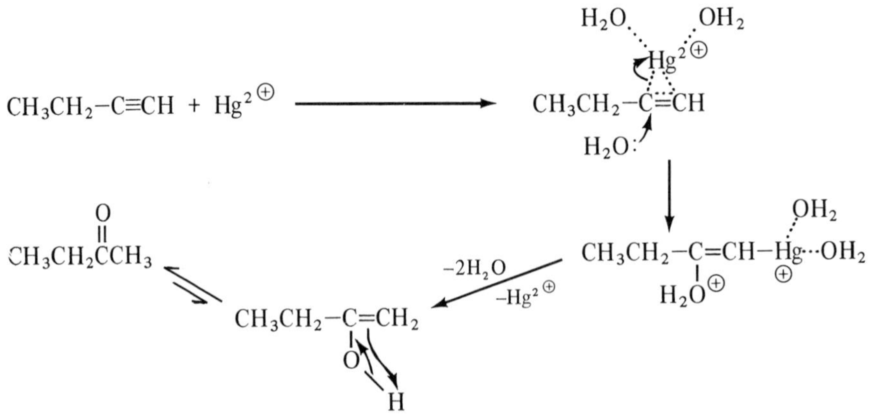

$$CH_3CH_2-C\equiv CH + Hg^{2\oplus} \longrightarrow CH_3CH_2-\overset{\overset{\displaystyle H_2O\cdots Hg^{2\oplus}\cdots OH_2}{|}}{C}=CH$$

with $H_2O:$ attacking

$$\longrightarrow CH_3CH_2\overset{O}{\overset{\|}{C}}CH_3 \xleftarrow{} CH_3CH_2-\underset{\overset{|}{O}\cdots H}{C}=CH_2 \xleftarrow[-Hg^{2\oplus}]{-2H_2O} CH_3CH_2-\underset{\overset{|}{H_2O^{\oplus}}}{C}=CH-Hg\cdots OH_2 \quad (OH_2)$$

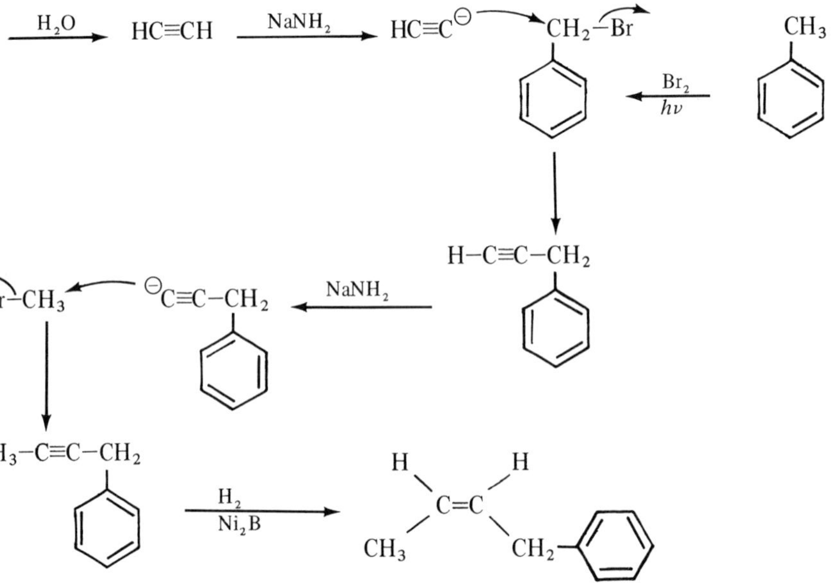

$$\xrightarrow{H_2O} HC\equiv CH \xrightarrow{NaNH_2} HC\equiv C^{\ominus} \curvearrowright CH_2-Br \xleftarrow{Br_2, h\nu} CH_3\text{-}(C_6H_5)$$

$$\longrightarrow H-C\equiv C-CH_2-(C_6H_5)$$

$$\xrightarrow{} Br-CH_3 \curvearrowleft {}^{\ominus}C\equiv C-CH_2-(C_6H_5) \xleftarrow{NaNH_2}$$

$$\longrightarrow CH_3-C\equiv C-CH_2-(C_6H_5) \xrightarrow{H_2, Ni_2B} \underset{CH_3}{\overset{H}{>}}C=C\underset{CH_2-(C_6H_5)}{\overset{H}{<}}$$

---

71. $CH_3CH_2CH_2C\equiv CH \xrightarrow{HCl} CH_3CH_2CH_2-\underset{}{\overset{Cl}{\underset{|}{C}}}=CH_2 \xrightarrow{HCl} CH_3CH_2CH_2\underset{\overset{|}{Cl}}{\overset{\overset{|}{Cl}}{C}}-CH_3$

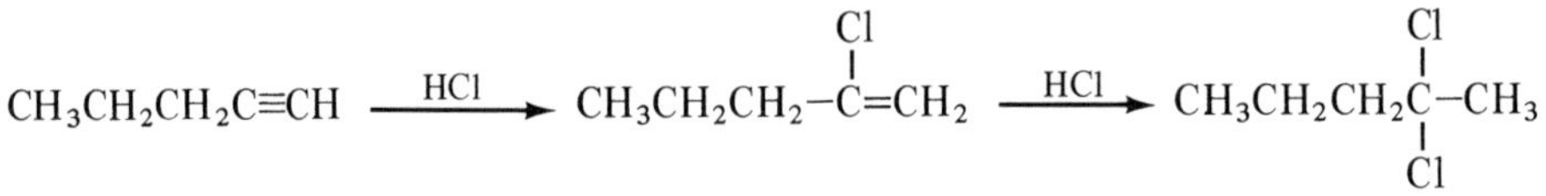

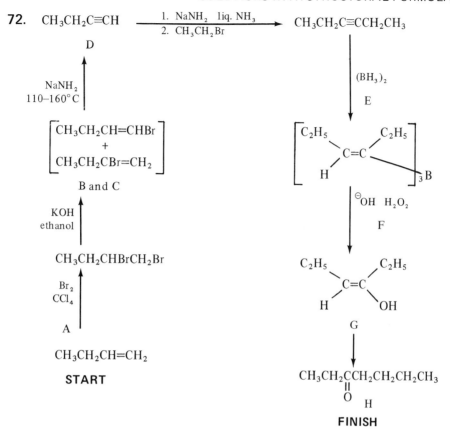

74. $CH_2=CH-CH=CH_2 + H-Br \longrightarrow \underset{Br^-}{CH_2=CH-\overset{\oplus}{CH}-CH_3} \longleftrightarrow \overset{\oplus}{CH_2}-CH=CH-CH_3$

fast ↓↑   slow ↓↑

$\underset{Br}{CH_2=CH-\overset{|}{CH}-CH_3}$ $\overset{Br}{CH_2}-CH=CH-CH_3$

Kinetic product favored at low temperature (least stable)   Thermodynamic product favored at high temperature (more stable)

75. $CH_3O_2C-CHBr-CH_2-CO_2CH_3 \xrightarrow{t\text{-BuO}^{\ominus}K^{\oplus}}$ [cis alkene with CH_3O_2C and CO_2CH_3] + [trans alkene with CO_2CH_3 and CO_2CH_3]

[Diels-Alder reactions of cyclohexadiene/benzene-type diene with dimethyl maleate and dimethyl fumarate, giving bicyclic adducts]

Major (endo)   +   Minor (exo)

76.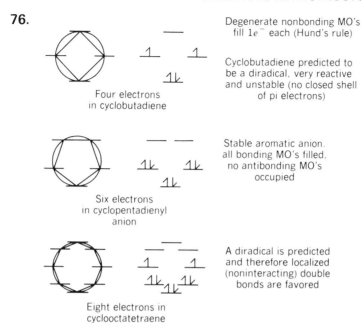

77.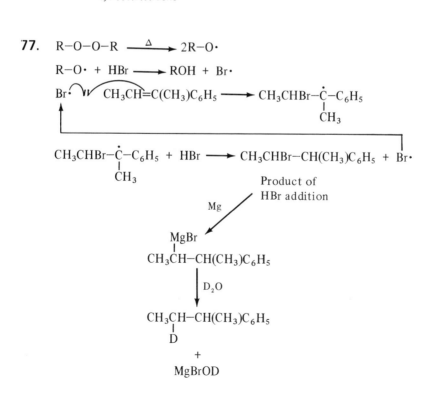

78. $(CH_3)_2CH-C_6H_5 \xrightarrow[Br_2]{h\nu} (CH_3)_2CBr-C_6H_5 \xrightarrow[Et_2O]{Mg} (CH_3)_2C(C_6H_5)-MgBr$ + CH_3CHO

X

$\downarrow$ work-up

$H^{\oplus}$ + HO-CH(CH_3)-C(CH_3)_2(C_6H_5)

$\longrightarrow [CH_3-\overset{\oplus}{C}-CH(CH_3)_2 \text{ (phenyl)}] \xrightarrow{-H^{\oplus}}$

A: $CH_3(C_6H_5)C=C(CH_3)_2$

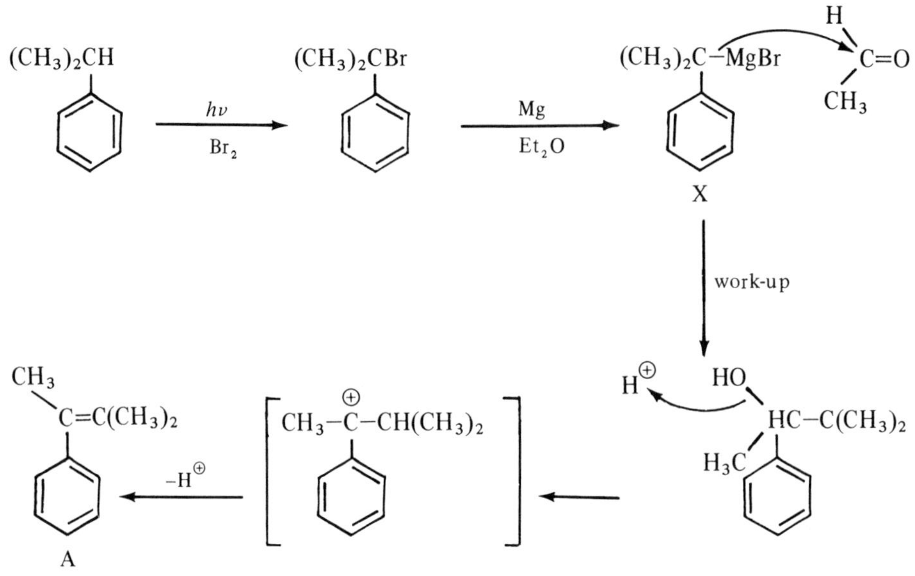

79. $CH_3-CHCl-CH_3 + AlCl_3 \rightleftharpoons CH_3-\overset{\oplus}{C}H-CH_3 + AlCl_4^{\ominus}$

The ortho position is hindered and attack there is less favored

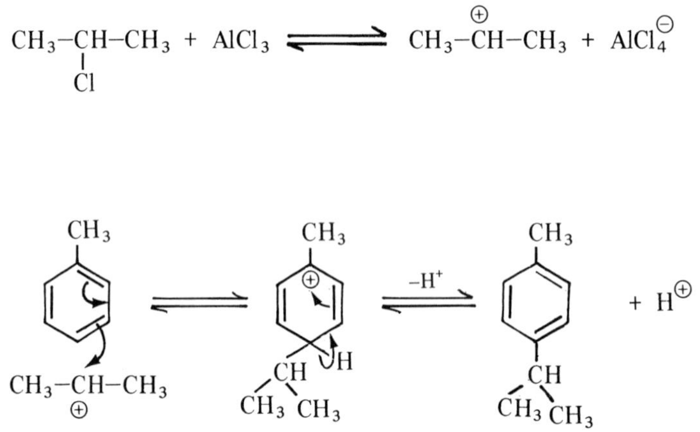

[mechanism showing toluene + isopropyl cation → arenium intermediate → p-isopropyltoluene + $H^{\oplus}$]

80. (c) benzene $\xrightarrow[AlCl_3]{CH_3Cl}$ toluene $\xrightarrow{KMnO_4}$ $\phi-CO_2H$ $\xrightarrow{Br_2/Fe}$ 3-bromobenzoic acid

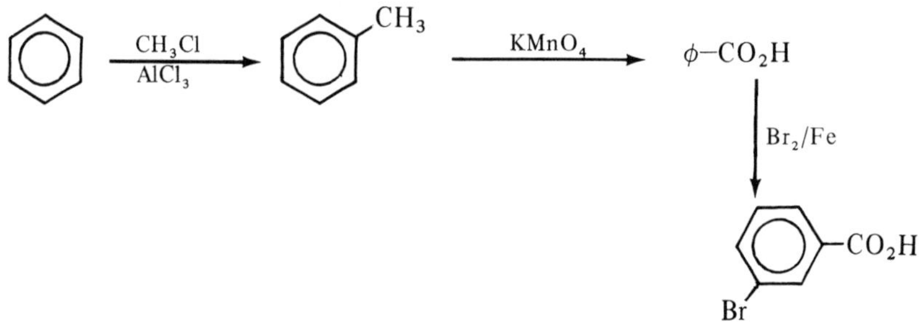

SOLUTIONS WITH STRUCTURAL FORMULAS 135

81.

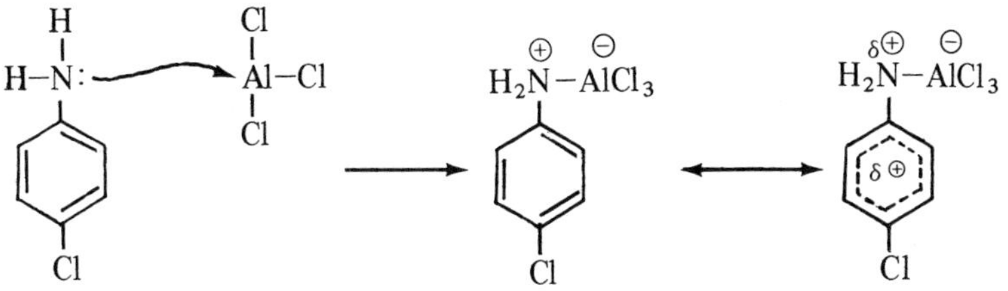

Ring deactivated towards electrophilic attack

Too unreactive to undergo Friedel–Crafts alkylation

82.

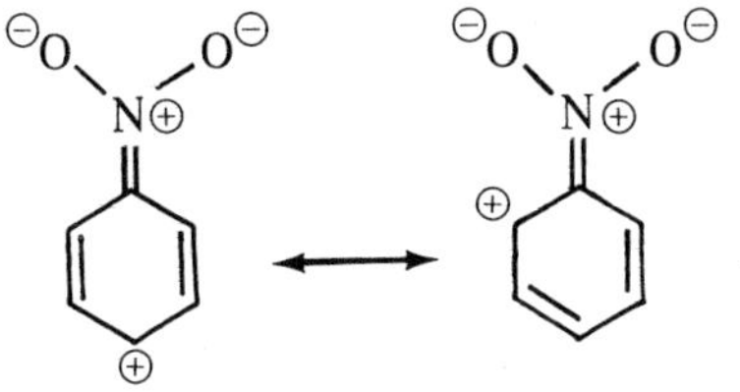

Partial positive charge on para position so attack at this position by $NO_2^{\oplus}$ is not favored

(ground state)

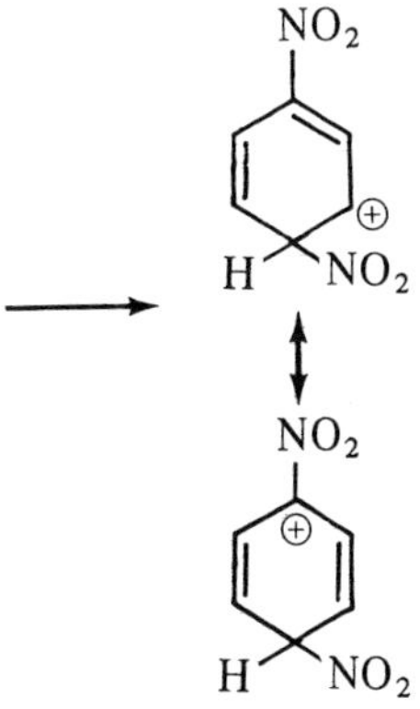

High energy intermediate because $\oplus$ charge is partially on carbon atom bonded to strongly electron-withdrawing nitro group

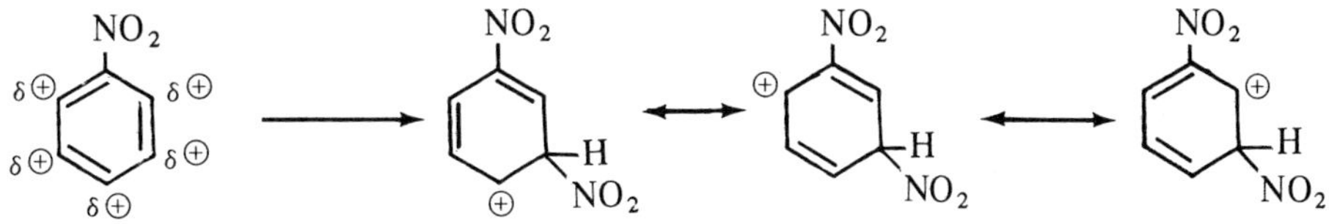

Lower energy intermediate compared to para substitution (above) as $\oplus$ charge is not placed on carbon atom bonded to a nitro group

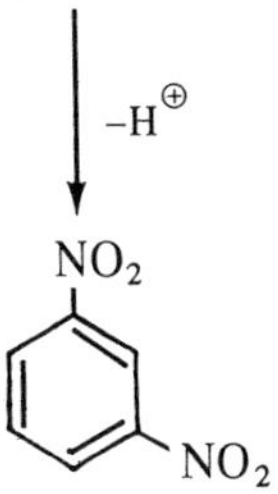

$\downarrow -H^{\oplus}$

Favored products

83.

84. Less important (2p-3p overlap)

More important (2p-2p overlap)

Major products are the 4-bromo- and 6-bromo-2-chloroanilines

85.

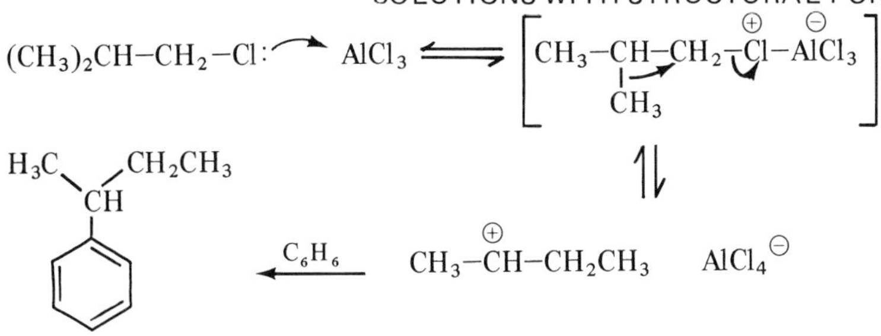

$(CH_3)_2CH-CH_2-Cl: \rightarrow AlCl_3 \rightleftharpoons \left[CH_3-\underset{CH_3}{\underset{|}{CH}}-CH_2 \overset{\oplus}{-} \overset{\ominus}{Cl}-AlCl_3 \right]$

$\updownarrow$

[phenyl ring with $H_3C\diagdown\diagup CH_2CH_3$ / CH substituent] $\xleftarrow{C_6H_6}$ $CH_3-\overset{\oplus}{CH}-CH_2CH_3 \quad AlCl_4^{\ominus}$

or alternatively

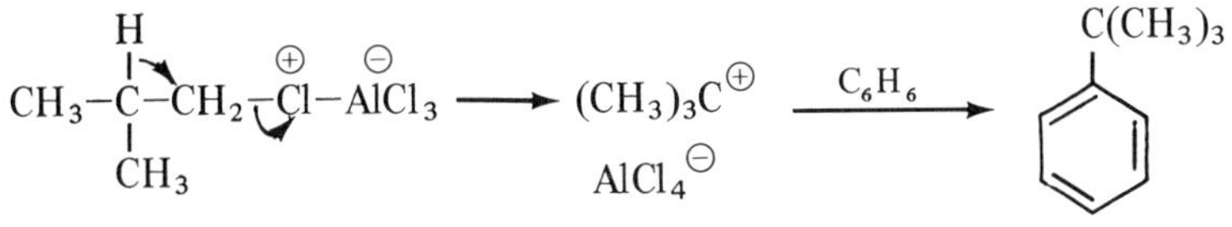

$CH_3-\underset{CH_3}{\overset{H}{\underset{|}{\overset{|}{C}}}}-CH_2 \overset{\oplus}{-} \overset{\ominus}{Cl}-AlCl_3 \longrightarrow (CH_3)_3C^{\oplus} \quad AlCl_4^{\ominus} \xrightarrow{C_6H_6}$ [phenyl ring with $C(CH_3)_3$]

86. [toluene] $\xrightarrow[\text{reflux}]{\substack{\text{conc.}\\ \text{HNO}_3\\ \text{fuming H}_2\text{SO}_4}}$ [2,4,6-trinitrotoluene: ring with CH_3, O_2N, NO_2, NO_2] $\xrightarrow[\ominus OH \; \Delta]{KMnO_4 \; H^{\oplus}}$ [2,4,6-trinitrobenzoic acid: ring with CO_2H, O_2N, NO_2, NO_2]

$\xrightarrow[-CO_2]{-H^{\oplus} \; \Delta}$

[1,3,5-trinitrobenzene: ring with O_2N, NO_2, NO_2] $\xleftarrow{H_2O}$ $\left[\text{ring with } \ominus, O_2N, NO_2, NO_2 \right]$ The negative charge in this intermediate is stabilized inductively by the nitro groups on the ring

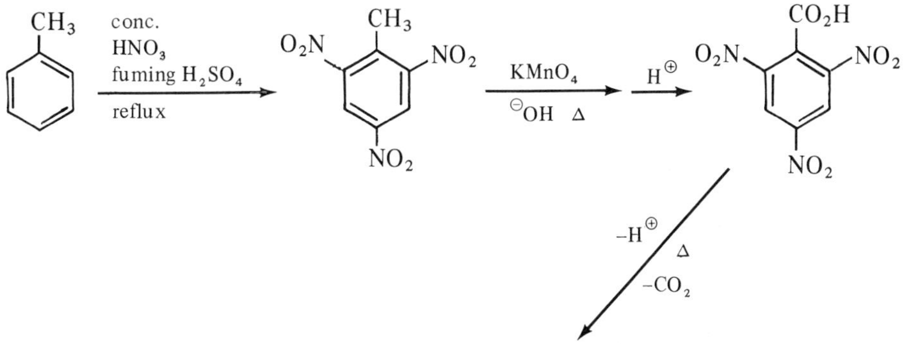

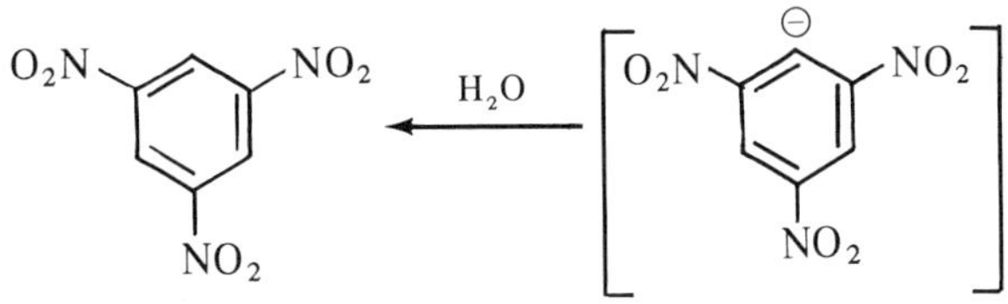

87.

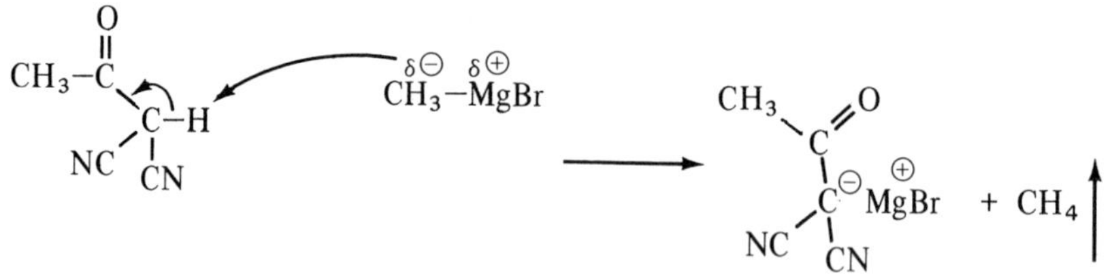

$$CH_3-\overset{O}{\overset{\|}{C}}-\underset{\underset{CN}{NC}}{\overset{}{C}}-H \quad \overset{\delta\ominus}{CH_3}-\overset{\delta\oplus}{MgBr} \longrightarrow CH_3-\overset{O}{\overset{\|}{C}}-\underset{\underset{CN}{NC}}{\overset{}{C}}{}^{\ominus} \cdot {}^{\oplus}MgBr + CH_4 \uparrow$$

88.

$$C_4H_9Br + NEt_3 \longrightarrow C_4H_9\overset{\oplus}{N}Et_3 \cdot \overset{\ominus}{Br} \quad (\overset{\oplus}{NR_4}, \overset{\ominus}{Br} \text{ below})$$

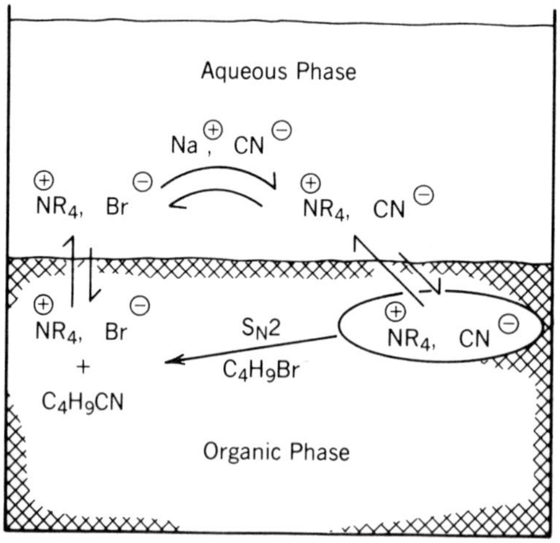

Aqueous Phase

$Na^{\oplus}, CN^{\ominus}$

$\overset{\oplus}{NR_4}, \overset{\ominus}{Br} \rightleftharpoons \overset{\oplus}{NR_4}, CN^{\ominus}$

$\overset{\oplus}{NR_4}, \overset{\ominus}{Br} \quad \underset{C_4H_9Br}{\overset{S_N2}{\longleftarrow}} \quad \overset{\oplus}{NR_4}, CN^{\ominus}$
$+$
C_4H_9CN

Organic Phase

89.

[Reaction scheme: 2-(2-chloro-3,5-dinitrophenyl)ethanol + ⁻OCH₃ ⇌ alkoxide intermediate → Meisenheimer complex (shown as two resonance structures) → 4,6-dinitro-2,3-dihydrobenzofuran + Cl⁻]

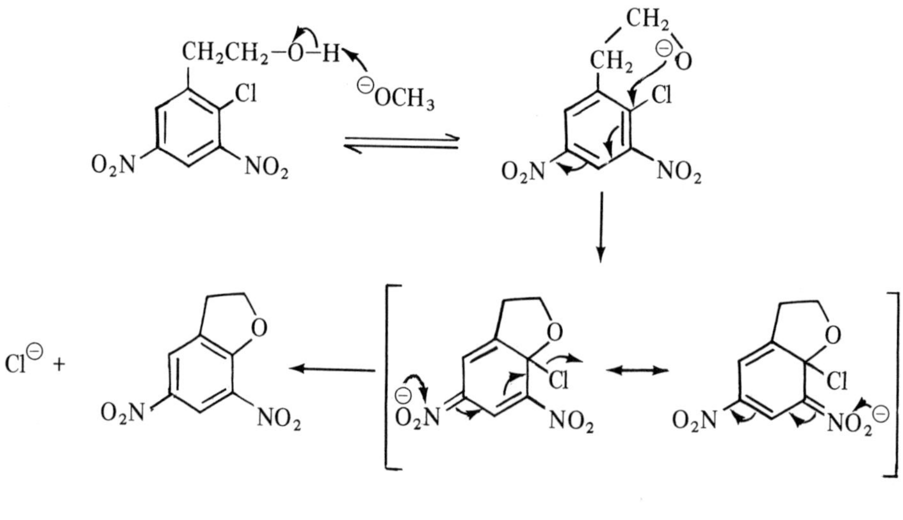

90. $C_6H_6 + CH_3CH_2Cl \xrightarrow{AlCl_3} C_6H_5-CH_2CH_3 \xrightarrow{Br_2, h\nu} C_6H_5-CHBr-CH_3$

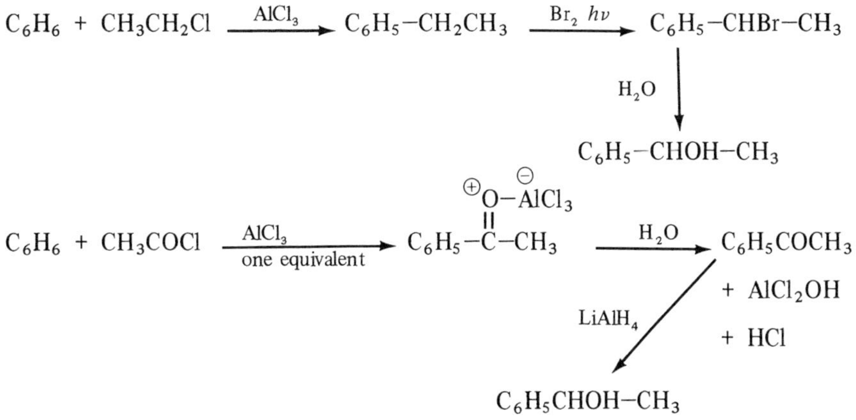

$\xrightarrow{H_2O} C_6H_5-CHOH-CH_3$

$C_6H_6 + CH_3COCl \xrightarrow[\text{one equivalent}]{AlCl_3} C_6H_5-\overset{\overset{\oplus}{O}-\overset{\ominus}{AlCl_3}}{\underset{\|}{C}}-CH_3 \xrightarrow{H_2O} C_6H_5COCH_3 + AlCl_2OH + HCl$

$\xrightarrow{LiAlH_4} C_6H_5CHOH-CH_3$

91.

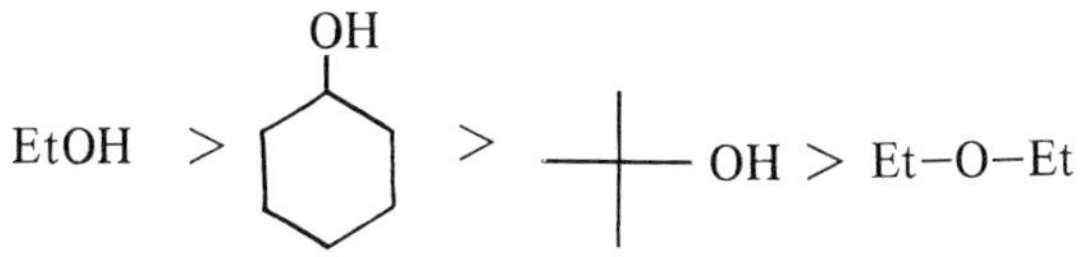

$EtOH > \underset{\text{cyclohexanol}}{\bigcirc\!\!-OH} > \underset{\text{t-butanol}}{\diagup\!\!\!-OH} > Et-O-Et$

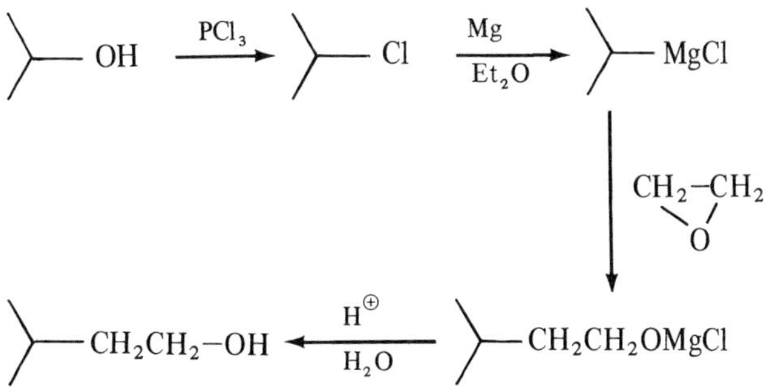

92.

$\diagup\!\!\!-OH \xrightarrow{PCl_3} \diagup\!\!\!-Cl \xrightarrow[Et_2O]{Mg} \diagup\!\!\!-MgCl$

$\xrightarrow{\overset{CH_2-CH_2}{\underset{O}{\diagdown\!/}}} \diagup\!\!\!-CH_2CH_2OMgCl \xrightarrow[H_2O]{H^\oplus} \diagup\!\!\!-CH_2CH_2-OH$

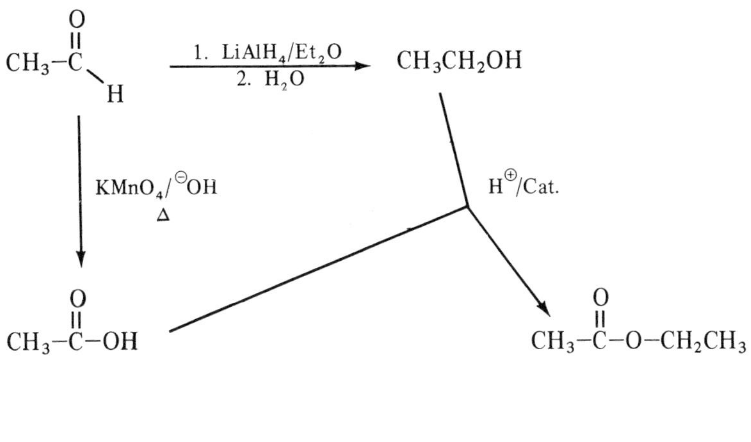

93. $CH_3-\underset{H}{\overset{\overset{O}{\|}}{C}} \xrightarrow[\text{2. }H_2O]{\text{1. LiAlH}_4/Et_2O} CH_3CH_2OH$

$\downarrow KMnO_4/{}^\ominus OH, \Delta \qquad \qquad \downarrow H^\oplus/Cat.$

$CH_3-\underset{\|}{\overset{O}{C}}-OH \qquad\qquad CH_3-\underset{\|}{\overset{O}{C}}-O-CH_2CH_3$

140 SOLUTIONS WITH STRUCTURAL FORMULAS

94.

[⁻O–C₆H₄–CH₂ ⟷ O=C₆H₄=CH₂ (Least stable anion) ⟷ C₆H₅–CH₂⁻]

↑
Very unlikely resonance form

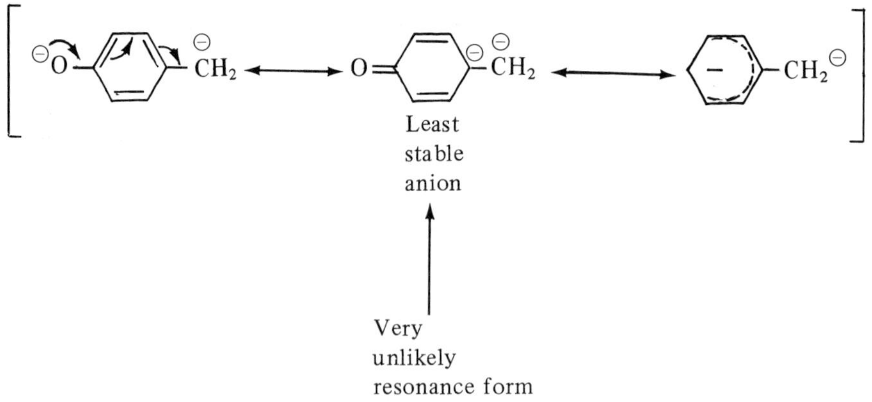

95. $Zn + 2HCl \longrightarrow ZnCl_2 + H_2$

$R-\ddot{O}-H + ZnCl_2 \rightleftharpoons R-\overset{\oplus}{\underset{H}{O}}-\overset{\ominus}{Zn}Cl_2$

$Cl^{\ominus} \curvearrowright R-\overset{\oplus}{\underset{H}{O}}-\overset{\ominus}{Zn}Cl_2 \longrightarrow R-Cl + [Zn(OH)Cl_2]^{\ominus}$

$[Zn(OH)Cl_2]^{\ominus} + H^{\oplus} \rightleftharpoons ZnCl_2 + H_2O$

96.

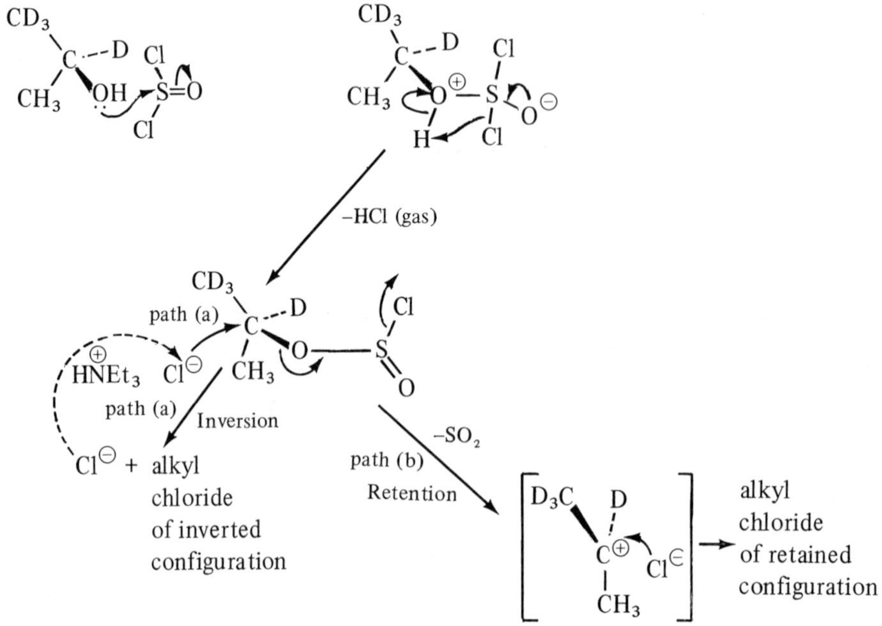

CD₃–C(D)(CH₃)–OH   +   Cl–S(=O)–Cl

→

CD₃–C(D)(CH₃)–O⁺(H)–S(Cl)(=O)–O⁻ ... Cl

–HCl (gas)

↓

path (a): CD₃–C(D)(CH₃)–O–S(Cl)=O
        HNEt₃⁺ Cl⁻
path (a) Inversion → Cl⁻ + alkyl chloride of inverted configuration

path (b) Retention –SO₂ → [D₃C–C⁺(D)(CH₃) Cl⁻] → alkyl chloride of retained configuration

SOLUTIONS WITH STRUCTURAL FORMULAS 141

97. (a) $\rangle\!\!-\!\!CH_2OH$

(b) $\rangle\!\!-\!\!CH_2CH_2OH$

(c) $\rangle\!\!-\!\!C(CH_3)_2OH$

(d) $\left(\rangle\!\!-\!\!\right)_2\!\!\underset{CH_3}{\overset{}{C}}\!\!-\!OH$

98. (see mechanism below)

50/50
A racemic form   (1)

$I^{\ominus}$ + (CH₃)(D)(H)C—Br $\xrightarrow{\text{inversion}}$ I—C(CH₃)(D)(H) + Br$^{\ominus}$ (2)

99.

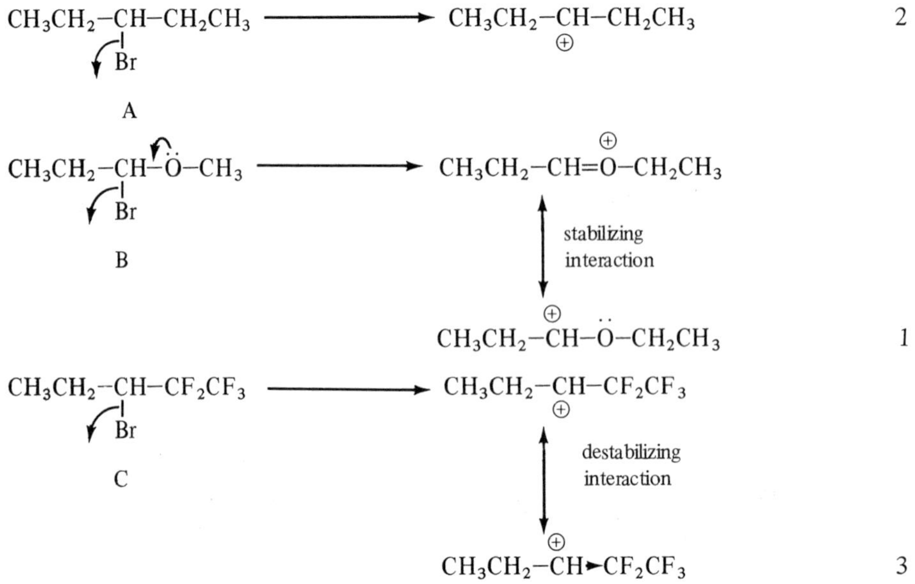

A: $CH_3CH_2-\underset{\underset{Br}{|}}{CH}-CH_2CH_3 \longrightarrow CH_3CH_2-\overset{\oplus}{CH}-CH_2CH_3$ **2**

B: $CH_3CH_2-\underset{\underset{Br}{|}}{CH}-\ddot{O}-CH_3 \longrightarrow CH_3CH_2-CH=\overset{\oplus}{O}-CH_2CH_3$

↕ stabilizing interaction

$CH_3CH_2-\overset{\oplus}{CH}-\ddot{O}-CH_2CH_3$ **1**

C: $CH_3CH_2-\underset{\underset{Br}{|}}{CH}-CF_2CF_3 \longrightarrow CH_3CH_2-\overset{\oplus}{CH}-CF_2CF_3$

↕ destabilizing interaction

$CH_3CH_2-\overset{\oplus}{CH}-CF_2CF_3$ **3**

100. The error was made in the choice of reagents. In choosing methoxide and *t*-butyl bromide the student introduced the possibility of elimination as follows:

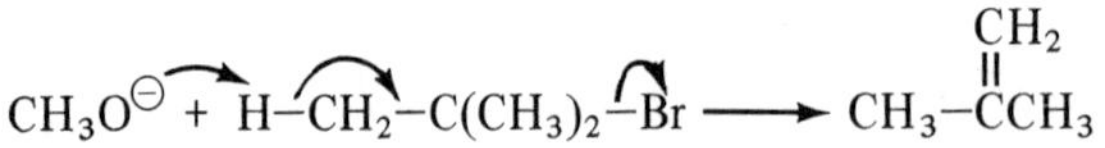

$CH_3O^\ominus + H-CH_2-C(CH_3)_2-Br \longrightarrow CH_3-\underset{\underset{CH_3}{|}}{\overset{\overset{CH_2}{\|}}{C}}CH_3$

However, if the student had chosen *t*-butoxide for a nucleophilic attack on bromomethane, elimination would not have been possible.

101. $C_6H_5-O-\underset{\underset{CH_3}{|}}{\overset{\overset{CH_3}{|}}{C}}-CH_3 \xrightarrow{HBr} C_6H_5-O-H + Br-\underset{\underset{CH_3}{|}}{\overset{\overset{CH_3}{|}}{C}}-CH_3$

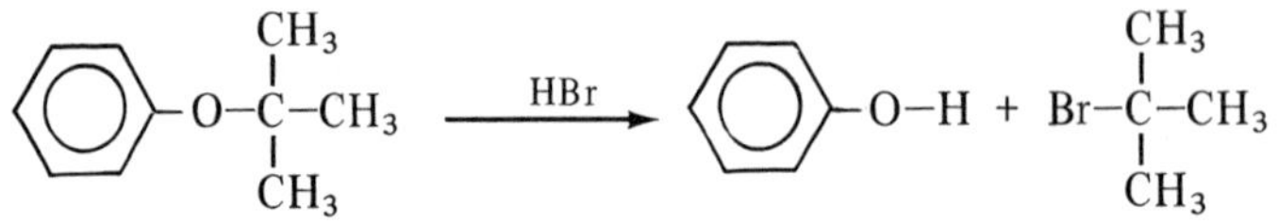

Mechanism

$C_6H_5-O-\underset{\underset{CH_3}{|}}{\overset{\overset{CH_3}{|}}{C}}-CH_3 \xrightarrow{HBr} C_6H_5-\overset{\oplus}{\underset{H}{O}}-\underset{\underset{CH_3}{|}}{\overset{\overset{CH_3}{|}}{C}}-CH_3$

↓

$CH_3-\underset{\underset{CH_3}{|}}{\overset{\overset{CH_3}{|}}{C}}-Br \xleftarrow{Br^\ominus} CH_3-\underset{\underset{CH_3}{|}}{\overset{\overset{CH_3}{|}}{\overset{\oplus}{C}}} + C_6H_5-OH$

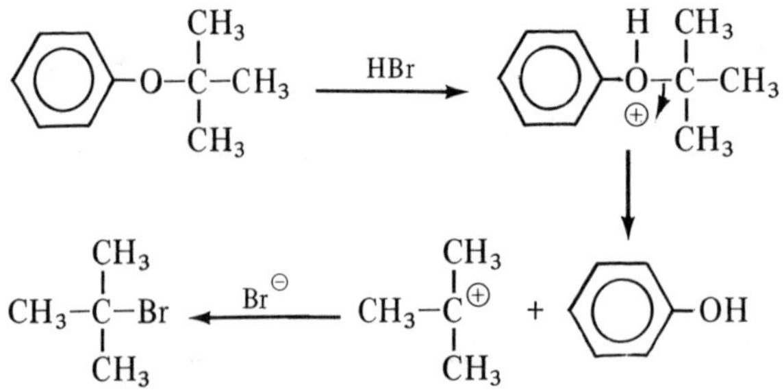

SOLUTIONS WITH STRUCTURAL FORMULAS 143

102. The 2,2,5,5-tetramethylcyclopentan-1-ol reacts with CH$_3^-$Li$^+$ in Et$_2$O to give the lithium alkoxide + CH$_4$↑. The alkoxide then attacks tosyl chloride (CH$_3$-C$_6$H$_4$-SO$_2$-Cl), displacing Cl$^-$, to give the cyclopentyl tosylate.

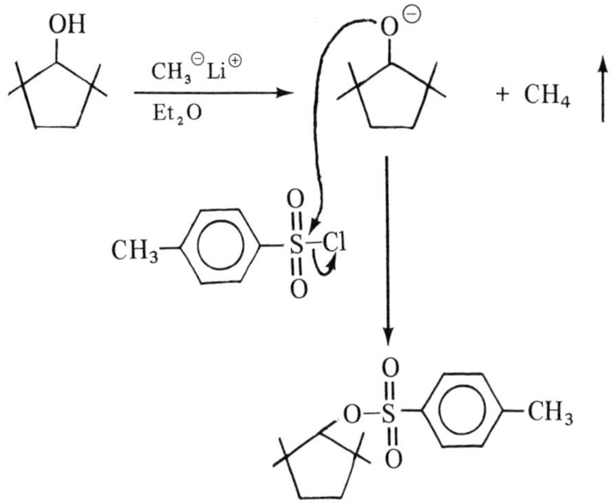

103.

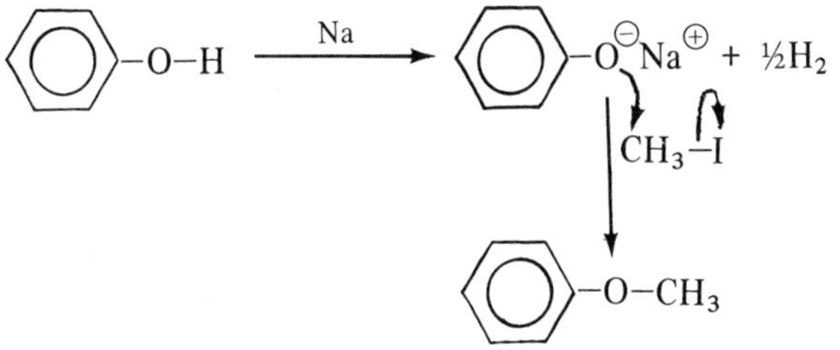

C$_6$H$_5$-O-H $\xrightarrow{\text{Na}}$ C$_6$H$_5$-O$^-$Na$^+$ + ½H$_2$

C$_6$H$_5$-O$^-$Na$^+$ + CH$_3$-I → C$_6$H$_5$-O-CH$_3$

Incorrect choice

CH$_3$OH $\xrightarrow{\text{Na}}$ CH$_3$O$^-$Na$^+$ $\xrightarrow{\text{C}_6\text{H}_5\text{-I}}$ CH$_3$O-C$_6$H$_5$

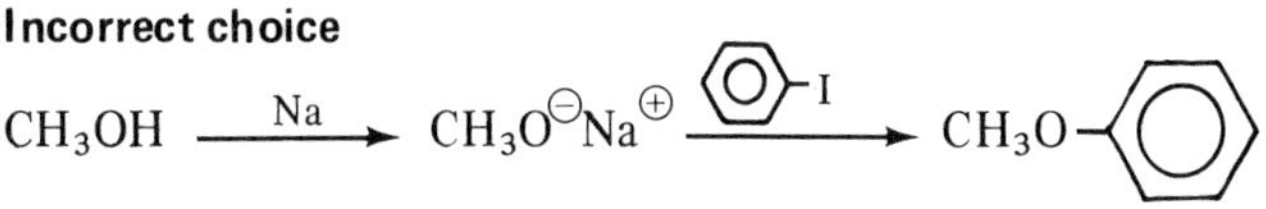

Does not form

104. HCCl$_3$ $\xrightarrow{\text{NaOH}}$ $^-$:CCl$_3$ $\xrightarrow{-\text{Cl}^-}$:CCl$_2$

Dichlorocarbene

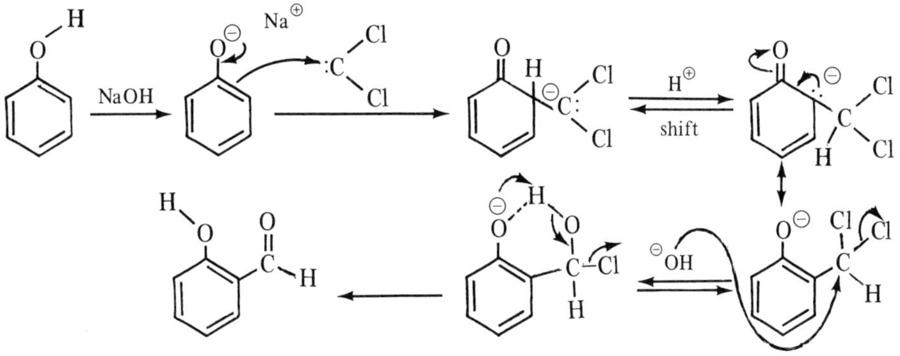

Phenol $\xrightarrow{\text{NaOH}}$ phenoxide (with Na$^+$) attacks :CCl$_2$ at ortho position → cyclohexadienone intermediate with -CCl$_2^-$ → H$^+$ shift → ortho-CHCl$_2$ phenoxide → $^-$OH adds, loses Cl$^-$ → gem-diol chloride → ortho-hydroxybenzaldehyde (salicylaldehyde).

105.

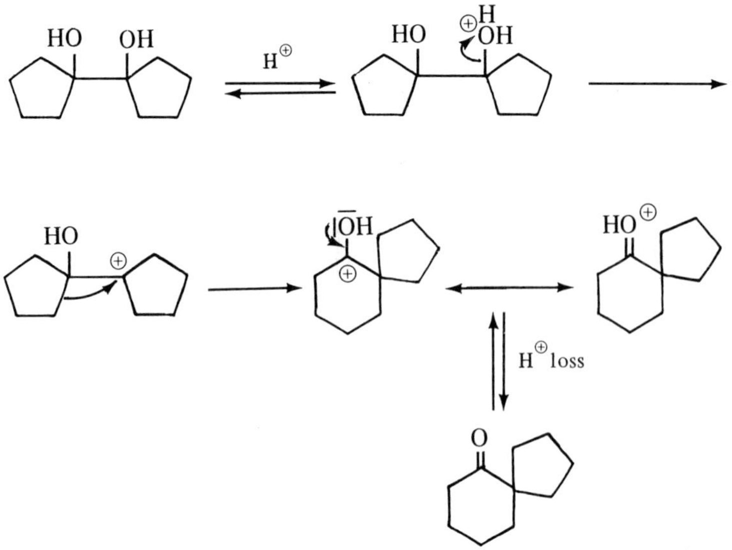
HO
OH
H⊕
HO
⊕H
OH
HO
⊕
OH
⊕
HO⊕
H⊕ loss
O

106.

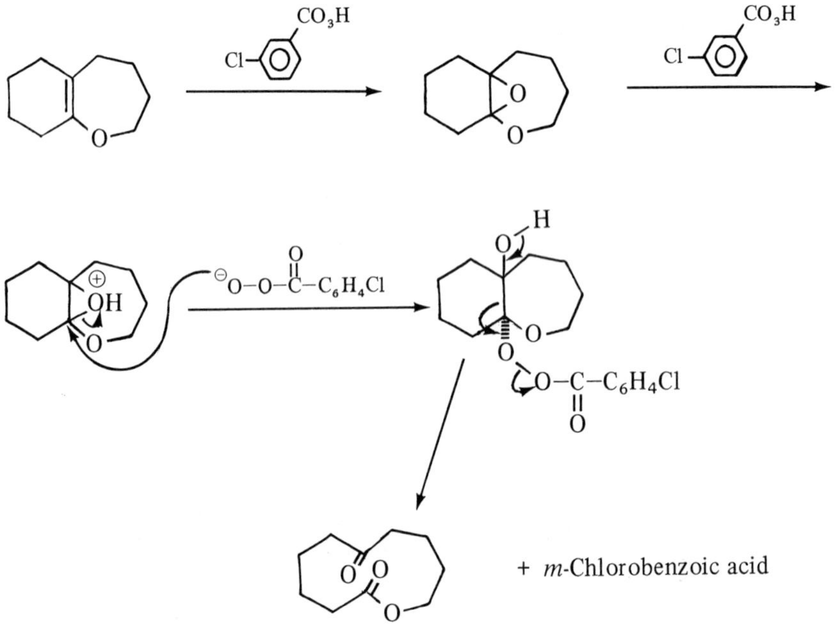
CO₃H
Cl
O
O
CO₃H
Cl
⊕
OH
O
⊖O–O–C–C₆H₄Cl
O
H
O
O
O
O–C–C₆H₄Cl
O
O
O
O
+ m-Chlorobenzoic acid

107.

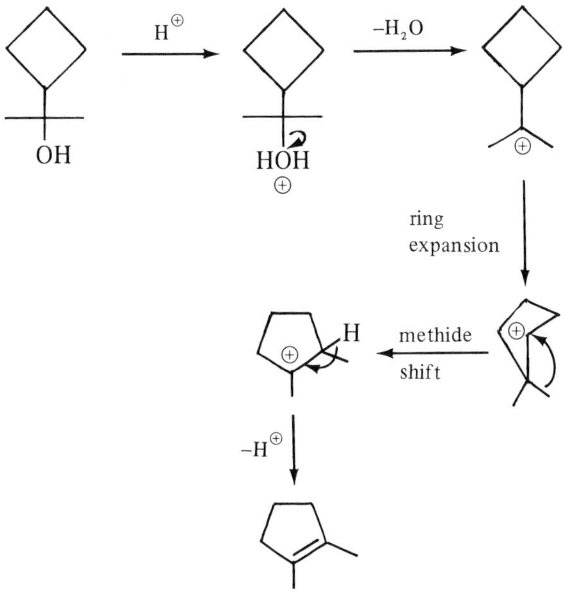
H⊕
–H2O
OH
HOH
ring
expansion
H
methide
shift
–H⊕

108. A > B

Compound A (*p*-nitrophenol) is more acidic because its conjugate base, *p*-nitrophenoxide ion, can be stabilized by the coplanar nitro group.

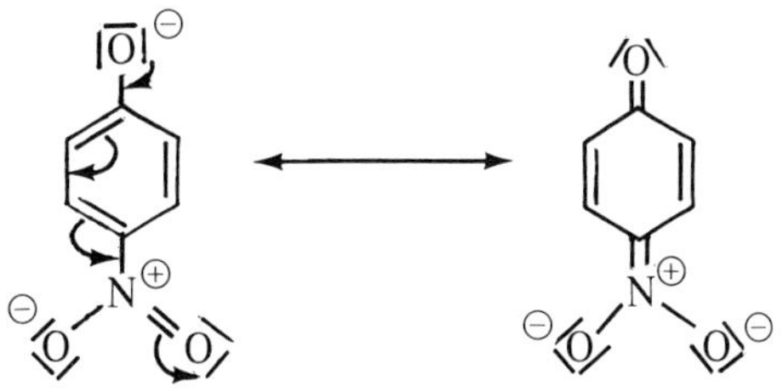

On the other hand, Compound B contains two bulky *t*-butyl groups that force the nitro group out of the ring plane. This inhibits resonance stabilization of the charge in the conjugate base. The inductive effect of the *t*-butyl groups decreases the acidity of Compound B, but not significantly. Recall that inductive effects drop off rapidly with distance.

109.

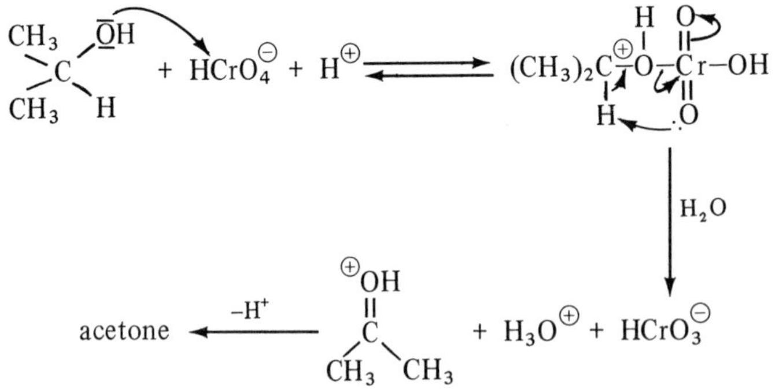

$$\underset{CH_3}{\overset{CH_3}{>}}\!C\!\underset{H}{\overset{\overline{O}H}{<}} + HCrO_4^{\ominus} + H^{\oplus} \rightleftharpoons (CH_3)_2C\!-\!O\!-\!\underset{\overset{\|}{O}}{\overset{H}{C}}r\!-\!OH$$

$$\downarrow H_2O$$

$$\text{acetone} \xleftarrow{-H^+} \underset{CH_3\ \ CH_3}{\overset{\overset{\oplus}{O}H}{\underset{\|}{C}}} + H_3O^{\oplus} + HCrO_3^{\ominus}$$

110.

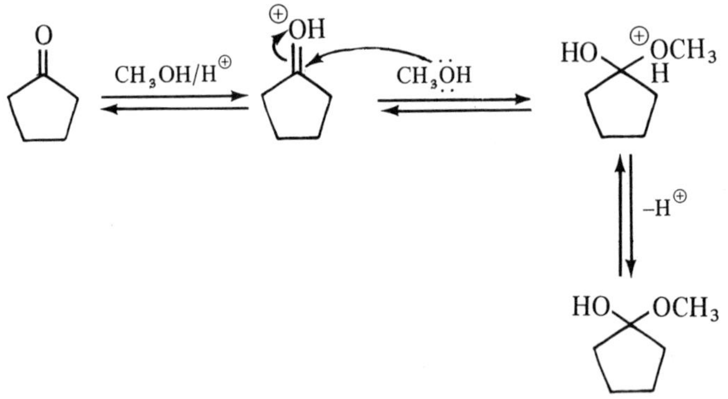

cyclopentanone $\xrightleftharpoons{CH_3OH/H^{\oplus}}$ [protonated ketone with $\overset{\oplus}{O}H$] $\xrightleftharpoons{CH_3\ddot{O}H}$ HO, $\overset{\oplus}{O}CH_3$, H

$$\updownarrow -H^{\oplus}$$

HO, OCH$_3$

111. $CH_2N_2 \equiv CH_2\!\!=\!\!\overset{\oplus}{N}\!\!=\!\!\overset{\ominus}{\underline{N}}| \longleftrightarrow \underset{}{\overset{\ominus}{C}H_2\!-\!\overset{\oplus}{N}\!\!\equiv\!\!N} \longleftrightarrow \underset{}{\overset{\ominus}{C}H_2\!-\!\underline{N}\!\!=\!\!\overset{\oplus}{N}}$

cyclohexanone + $\overset{\ominus}{C}H_2\!-\!\overset{\oplus}{N}\!\!\equiv\!\!N| \longrightarrow \left[\overset{\ominus}{O}\!-\!C(CH_2\!-\!\overset{\oplus}{N}\!\!\equiv\!\!N|) \longleftrightarrow \overset{\ominus}{O}\!-\!C(CH_2\!-\!\underline{N}\!\!=\!\!\overset{\oplus}{N}|) \right]$

$\overset{\ominus}{O}\!-\!C(CH_2\!-\!\overset{\oplus}{N}\!\!\equiv\!\!N|) \longrightarrow$ cycloheptanone + $|N\!\!\equiv\!\!N|$

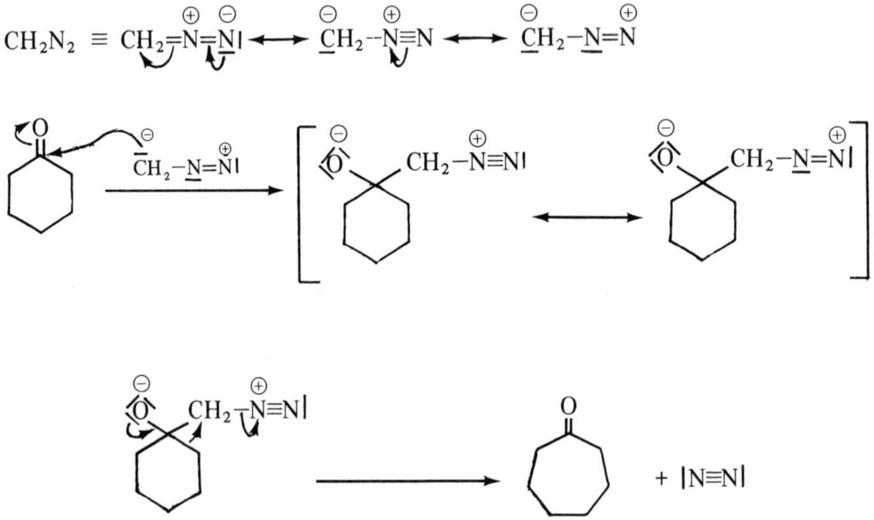

SOLUTIONS WITH STRUCTURAL FORMULAS

112.

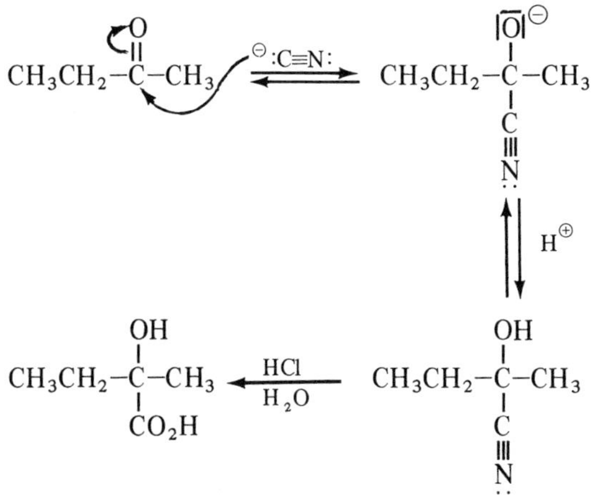

$$CH_3CH_2-\overset{\overset{O}{\|}}{C}-CH_3 \xrightarrow{\ominus :C\equiv N:} CH_3CH_2-\underset{\underset{\underset{\ddot{N}:}{\|\!|}}{C}}{\overset{\overset{|\overline{O}|^{\ominus}}{|}}{C}}-CH_3$$

$$\updownarrow H^{\oplus}$$

$$CH_3CH_2-\underset{\underset{CO_2H}{|}}{\overset{\overset{OH}{|}}{C}}-CH_3 \xleftarrow[H_2O]{HCl} CH_3CH_2-\underset{\underset{\underset{\ddot{N}:}{\|\!|}}{C}}{\overset{\overset{OH}{|}}{C}}-CH_3$$

113.

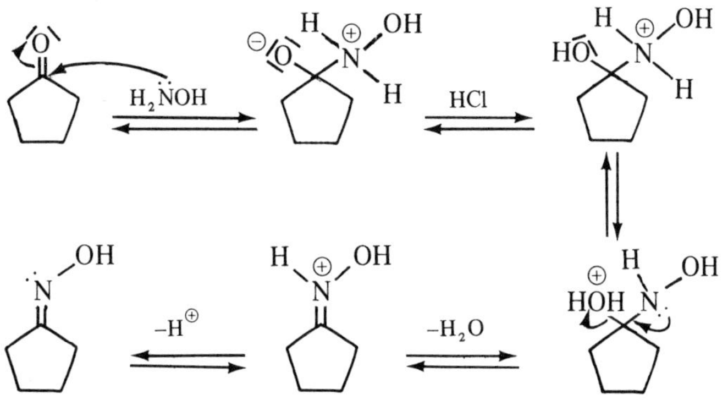

(mechanism for oxime formation from cyclopentanone + H_2NOH, then HCl, $-H_2O$, $-H^{\oplus}$)

114.

(a) 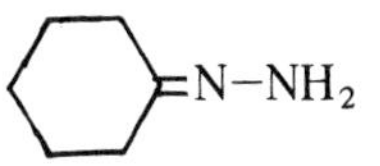 cyclohexanone =N−NH$_2$ a hydrazone

(b) 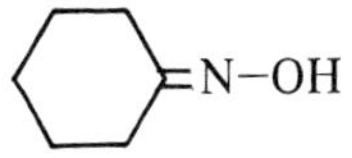 =N−OH   an oxime

(c) 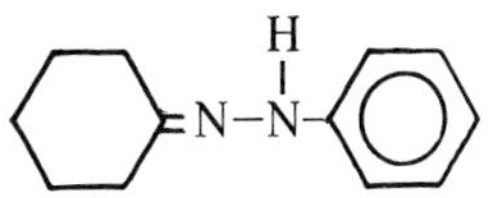 =N−N(H)−C$_6$H$_5$ a phenylhydrazone

(d) 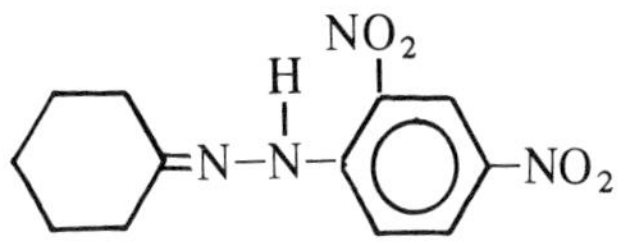 =N−N(H)−C$_6$H$_3$(NO$_2$)$_2$ a 2,4-dinitrophenylhydrazone

148 SOLUTIONS WITH STRUCTURAL FORMULAS

115.

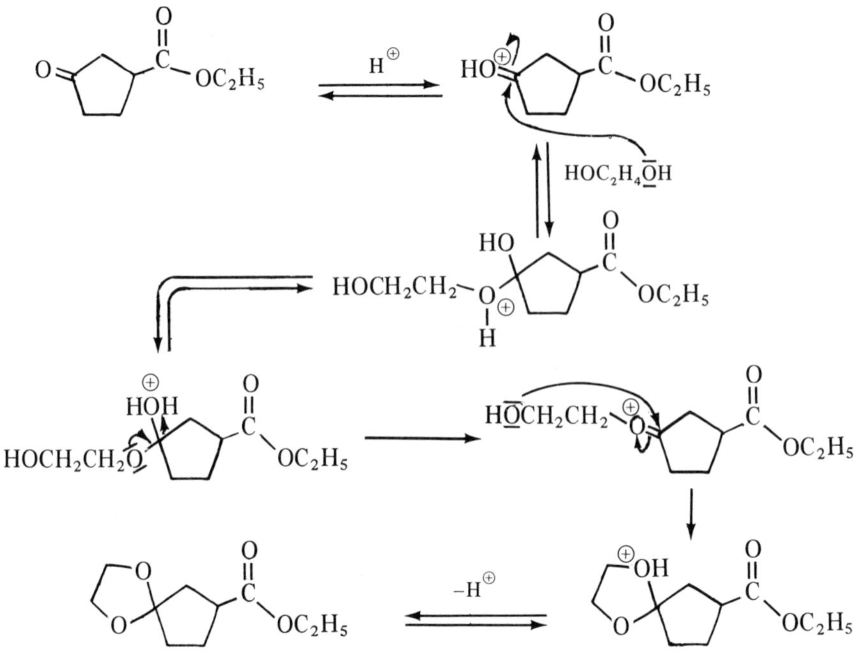

116.

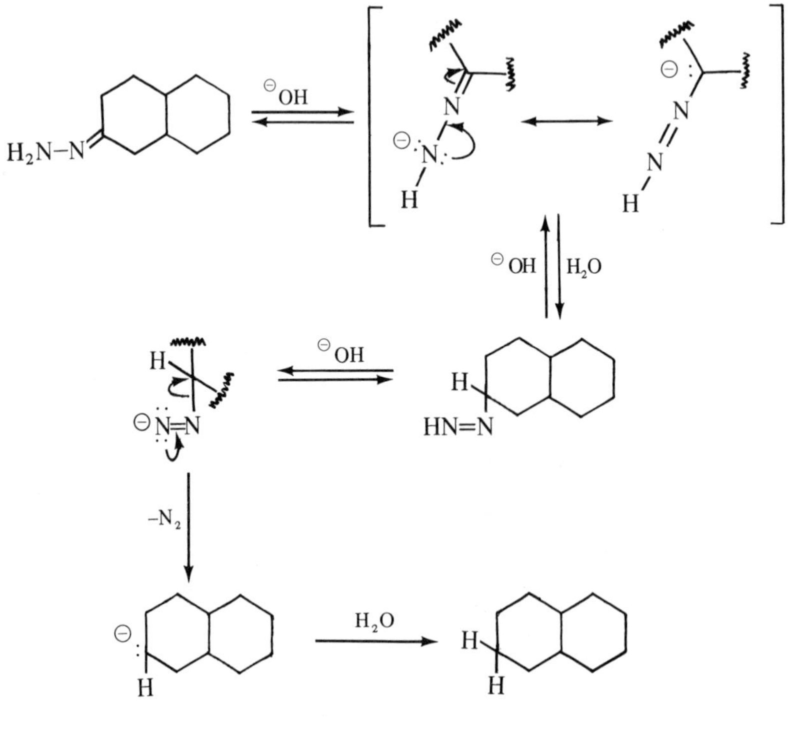

117.

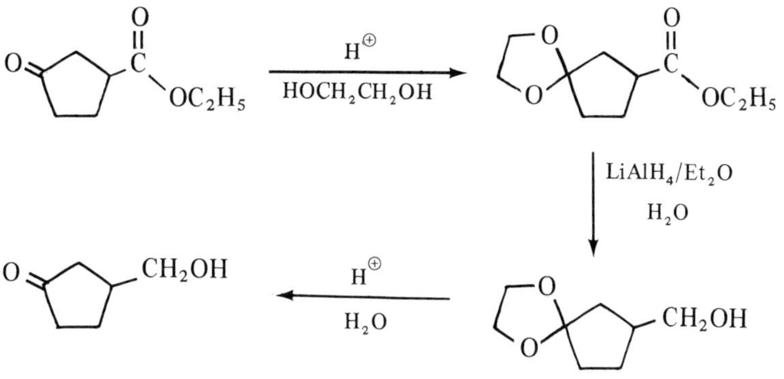

[Structural reaction scheme: ethyl 4-oxocyclopentanecarboxylate reacts with HOCH₂CH₂OH / H⁺ to form the ethylene ketal (ethyl ester with dioxolane ring). LiAlH₄/Et₂O, then H₂O converts the ester to CH₂OH. Then H⁺/H₂O hydrolyzes the ketal back to the ketone, giving 3-(hydroxymethyl)cyclopentanone.]

118. In the absence of the catalyst, CuCl, the Grignard reagent reacts to give (almost exclusively) products of 1,2-addition. When the ketone is added to a solution of cuprous chloride, however, addition is almost entirely 1,4. The effect of cuprous chloride is to produce the following intermediate:

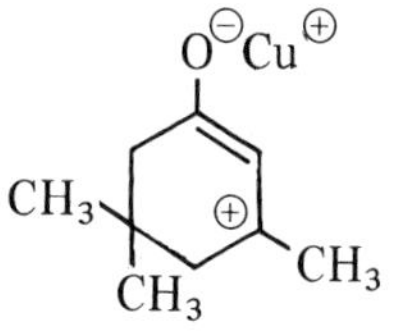

[Structure: cyclohexenone-derived enolate with O⁻Cu⁺, with CH₃ groups and a positive center on the ring]

Attack by the electronegative carbon atom of the Grignard can thus better occur at the tertiary center.

119.

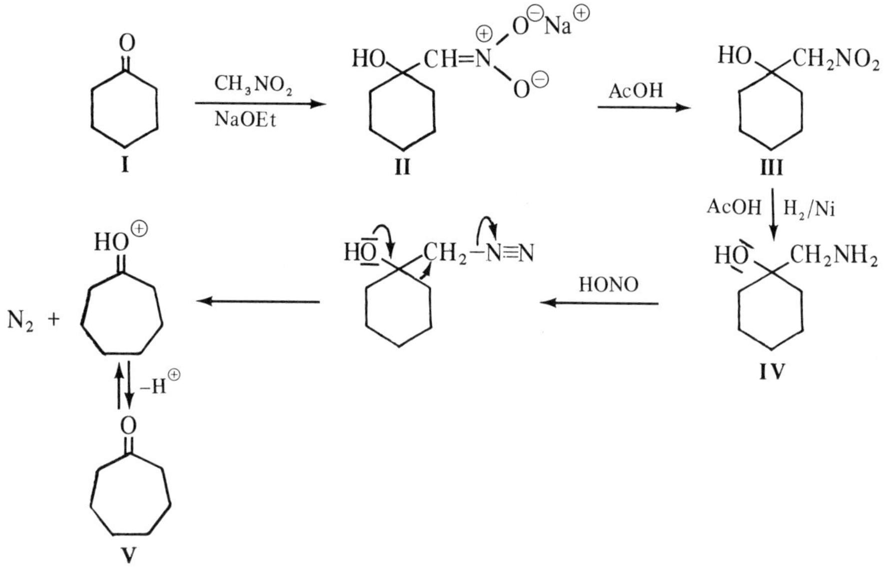

[Reaction scheme:
I: cyclohexanone
→ CH₃NO₂ / NaOEt →
II: 1-hydroxy-1-(CH=N(O⁻Na⁺)(O⁻))cyclohexane
→ AcOH →
III: 1-hydroxy-1-(CH₂NO₂)cyclohexane
→ AcOH, H₂/Ni →
IV: 1-hydroxy-1-(CH₂NH₂)cyclohexane
→ HONO →
HO–C(cyclohexane)–CH₂–N≡N (with arrows showing migration)
→ N₂ + protonated cycloheptanone (HO⁺=)
⇌ (−H⁺)
V: cycloheptanone]

120.

121.

152 SOLUTIONS WITH STRUCTURAL FORMULAS

122.

SOLUTIONS WITH STRUCTURAL FORMULAS   153

124. (a) 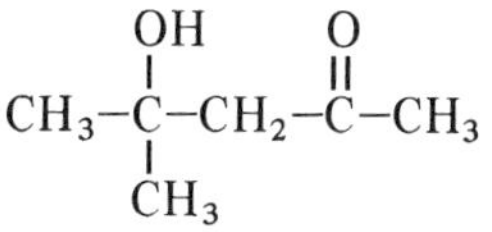 $CH_3-\underset{\underset{CH_3}{|}}{\overset{\overset{OH}{|}}{C}}-CH_2-\overset{\overset{O}{\|}}{C}-CH_3$

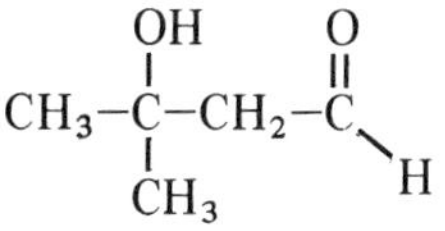

(b) $CH_3-\underset{\underset{CH_3}{|}}{\overset{\overset{OH}{|}}{C}}-CH_2-\overset{\overset{O}{\|}}{C}-H$

(c) $CH_3-\underset{\underset{H}{|}}{\overset{\overset{OH}{|}}{C}}-CH_2-\overset{\overset{O}{\|}}{C}-H$

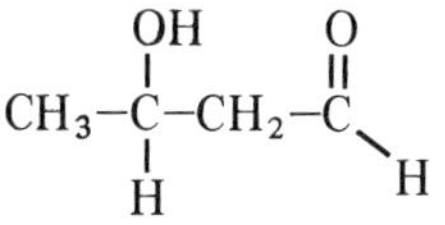

125. 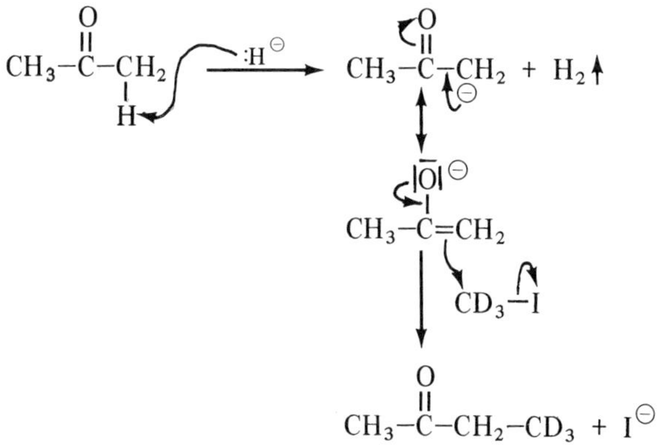 $CH_3-\overset{\overset{O}{\|}}{C}-CH_2 \overset{:H^\ominus}{\longrightarrow} CH_3-\overset{\overset{O}{\|}}{C}-CH_2^\ominus + H_2\uparrow$

$\updownarrow$

$CH_3-\underset{}{\overset{\overset{|\overline{O}|^\ominus}{|}}{C}}=CH_2$

$\downarrow CD_3-I$

$\updownarrow$

$CH_3-\overset{\overset{O}{\|}}{C}-CH_2-CD_3 + I^\ominus$

126. 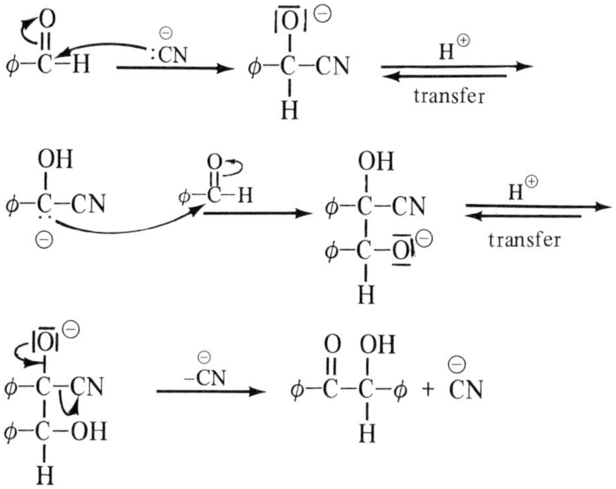 $\phi-\overset{\overset{O}{\|}}{C}-H \overset{:CN^\ominus}{\longrightarrow} \phi-\underset{H}{\overset{\overset{|\overline{O}|^\ominus}{|}}{C}}-CN \underset{transfer}{\overset{H^\oplus}{\rightleftharpoons}}$

$\phi-\underset{\ominus}{\overset{\overset{OH}{|}}{C}}-CN \quad \overset{\overset{O}{\|}}{C}-H \longrightarrow \underset{\underset{H}{|}}{\overset{\overset{OH}{|}}{\underset{\phi-C-\overline{O}|^\ominus}{\phi-C-CN}}} \underset{transfer}{\overset{H^\oplus}{\rightleftharpoons}}$

$\underset{\underset{H}{|}}{\overset{\overset{|\overline{O}|^\ominus}{|}}{\underset{\phi-C-OH}{\phi-C-CN}}} \overset{-CN}{\longrightarrow} \phi-\overset{\overset{O}{\|}}{C}-\underset{H}{\overset{\overset{OH}{|}}{C}}-\phi + CN^\ominus$

Benzoin

127. Intramolecular aldol condensation:

$$CH_3-\overset{O}{\underset{\|}{C}}-CH_2CH_2CH_2CH_2\overset{O}{\underset{\|}{CH}} \xrightarrow{\ominus OH} CH_3-\overset{O}{\underset{\|}{C}}-\underset{\ominus}{CH}CH_2CH_2CH_2\overset{O}{\underset{\|}{CH}} \longrightarrow$$

cyclopentane intermediate with $-\overset{\ominus}{O}$, H, and $\overset{O}{\underset{\|}{C}}-CH_3$ groups $\xrightleftharpoons{H_2O}$ cyclopentane with OH, H, and $\overset{O}{\underset{\|}{C}}-CH_3$ (+ $\ominus OH$) $\xrightarrow{-H_2O}$ cyclopentenyl-$\overset{O}{\underset{\|}{C}}-CH_3$ + H_2O

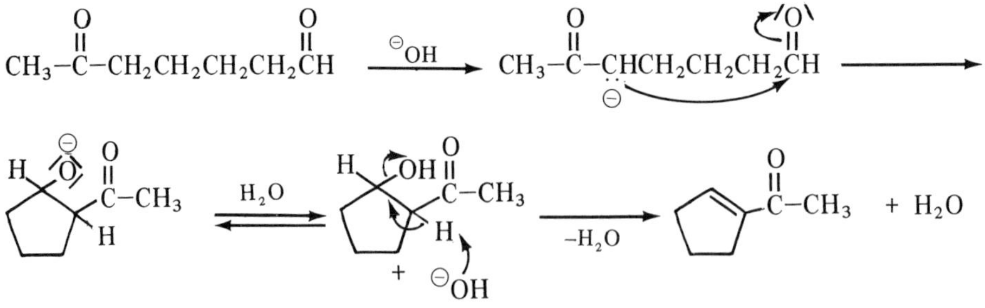

128. Retro-aldol:

$$CH_3-\underset{\underset{CH_3}{|}}{\overset{OH}{\underset{|}{C}}}-CH_2-\overset{O}{\underset{\|}{C}}-CH_3 \xrightleftharpoons{\ominus OH} CH_3-\underset{\underset{CH_3}{|}}{\overset{\ominus \overline{O}\overline{I}}{\underset{|}{C}}}-CH_2-\overset{O}{\underset{\|}{C}}-CH_3 \rightleftharpoons$$

$$CH_3-\overset{O}{\underset{\|}{C}}-CH_3 + \overset{\ominus}{CH_2}-\overset{O}{\underset{\|}{C}}-CH_3 \xrightleftharpoons{H_2O} 2\,CH_3-\overset{O}{\underset{\|}{C}}-CH_3 + \ominus OH$$

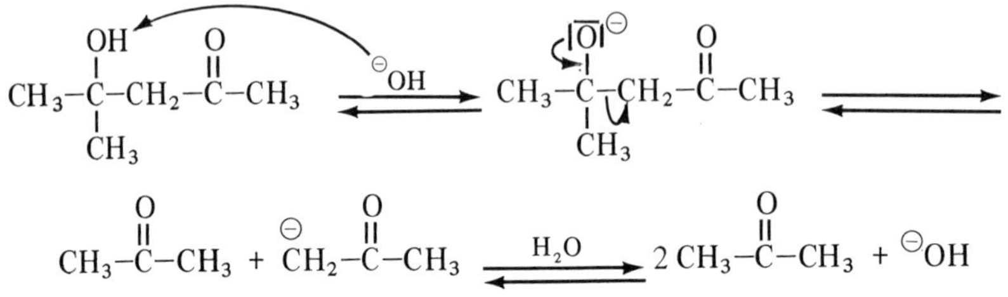

129. Crossed Cannizzaro reaction:

HCHO + $\overline{I}\overline{O}H^{\ominus}$ ⇌ H–C(H)(O⁻)(OH) → transfers hydride to 3-methoxybenzaldehyde (ArCHO where Ar = 3-methoxyphenyl)

→ 3-methoxybenzyl alkoxide (H–CH(O⁻)–C₆H₄–OCH₃) + HCOOH

→ 3-methoxybenzyl alcohol (CH₂OH–C₆H₄–OCH₃) + HCOO⁻

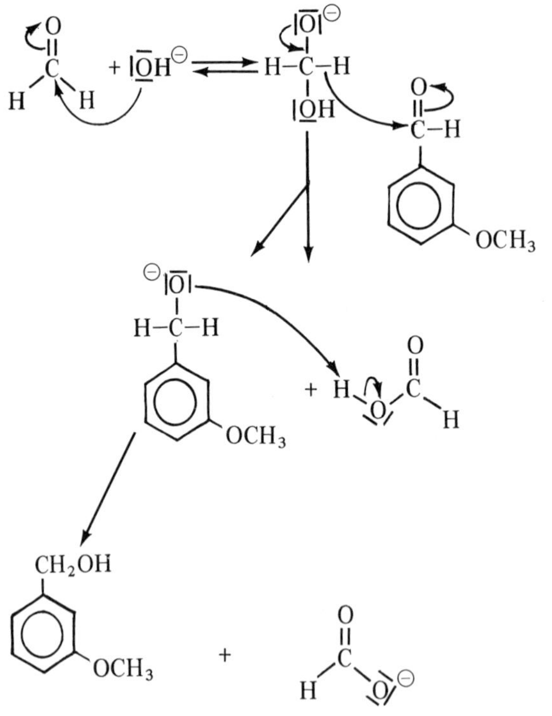

130.

132.

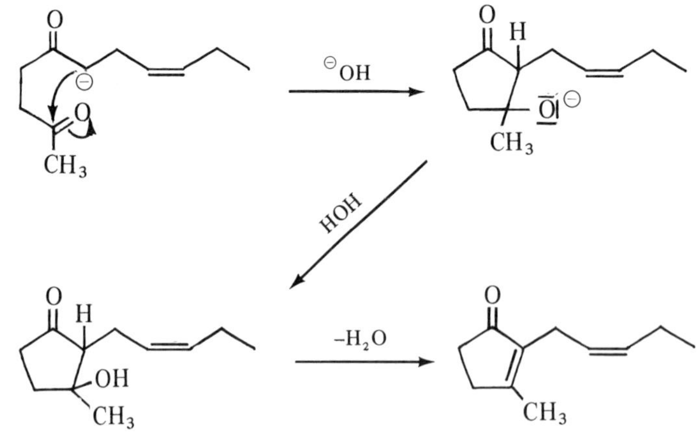

133.

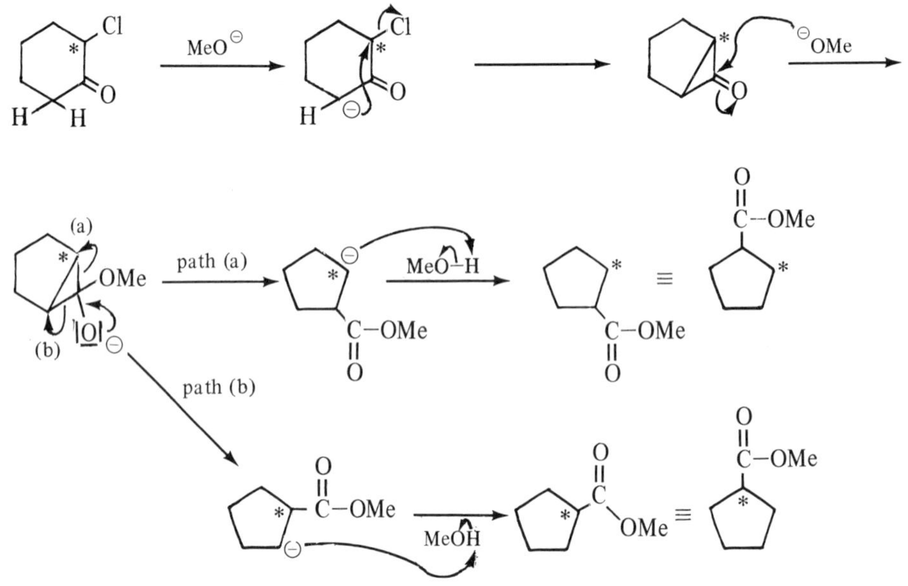

An alternative mechanism appears interesting until one examines the labeling information.

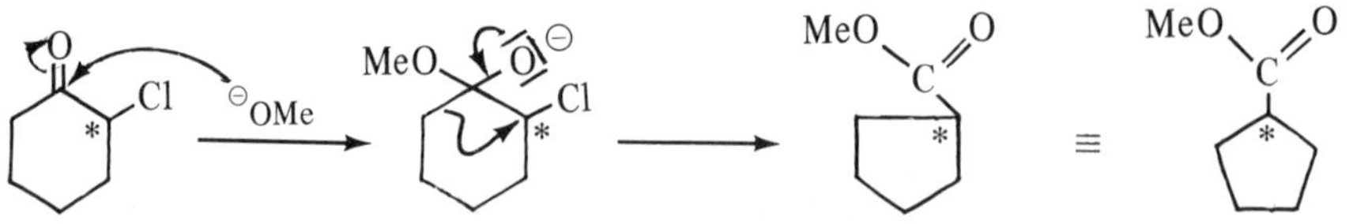

Only one product can result from this alternative route.

SOLUTIONS WITH STRUCTURAL FORMULAS 157

134.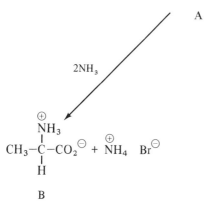

135. Products

(a) $CH_3-\underset{NH_2}{\underset{|}{\overset{O}{\overset{\|}{C}}}}$ + HCl

(b) cyclohexyl-$\overset{O}{\overset{\|}{C}}$-OEt + HCl

(c) phenyl-$\overset{O}{\overset{\|}{C}}$-N(H)CH$_2CH_3$ + HCl

(d) $CH_3-\overset{O}{\overset{\|}{C}}-O-\overset{O}{\overset{\|}{C}}-CH_3$ + NaCl

136. Products

(a) $2CH_3-\underset{\underset{OH}{}}{\overset{\overset{O}{\|}}{C}}$

(b) $CH_3-\underset{\underset{OH}{}}{\overset{\overset{O}{\|}}{C}}$ + HCl

(c) $CH_3-\underset{\underset{OH}{}}{\overset{\overset{O}{\|}}{C}}$ + NH_3

(d) $CH_3-\underset{\underset{OH}{}}{\overset{\overset{O}{\|}}{C}}$ + CH_3OH

137.

α-Naphthol + phosgene (Cl–CO–Cl) → naphthyl chloroformate + HCl; then with CH_3-NH_2 → Carbaryl (1-naphthyl N-methylcarbamate) + HCl

138.

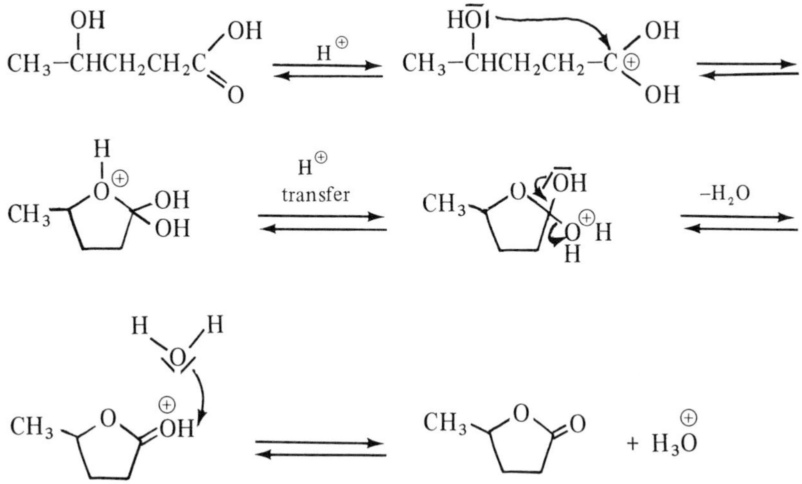

[Reaction scheme showing acid-catalyzed lactone formation from a γ-hydroxy acid: CH₃-CH(OH)CH₂CH₂C(=O)OH protonated, then cyclized through tetrahedral intermediates with proton transfers and loss of water to give the five-membered lactone (γ-butyrolactone derivative) + H₃O⁺.]

The β-hydroxy acid does not form a lactone because of the increased strain encountered in a four-membered ring relative to a five-membered ring. Instead of cyclizing to form the four-membered ring the β-hydroxybutyric acid dehydrates to form a double bond in conjugation with the carbonyl of the acid as follows:

$$CH_3-\underset{\underset{OH}{|}}{C}H-CH_2-CO_2H \xrightarrow{H^{\oplus}} CH_3CH=CH-CO_2H$$

139.

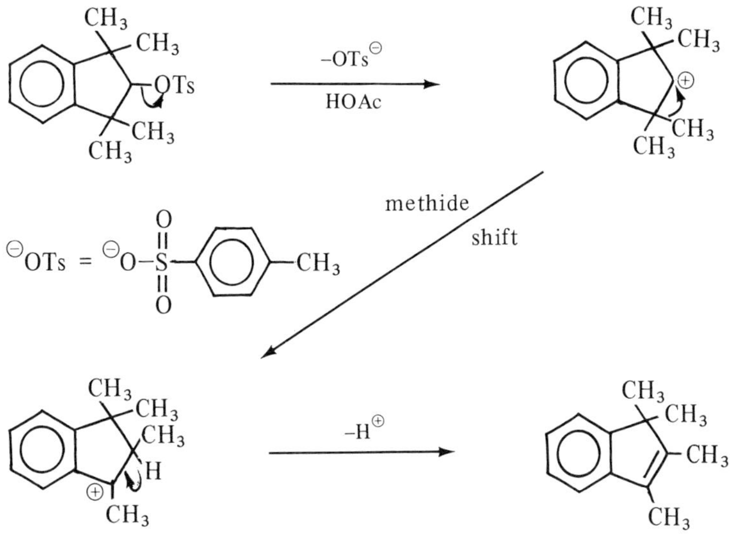

[Reaction scheme: 1,1,3,3-tetramethyl-2-indanyl tosylate ionizes (−OTs⁻, HOAc) to a secondary carbocation, which undergoes a methide shift to a more stable tertiary carbocation, then loss of H⁺ gives 1,1,2,3-tetramethyl-1H-indene.]

$$^{\ominus}OTs = {}^{\ominus}O-\underset{\underset{O}{\|}}{\overset{\overset{O}{\|}}{S}}-\!\!\left\langle\;\right\rangle\!\!-CH_3$$

160 SOLUTIONS WITH STRUCTURAL FORMULAS

140. (a) $\phi-CH_2\overset{O}{\underset{\underset{OAg}{|}}{C}} + Br_2 \xrightarrow{CCl_4} \phi-CH_2\overset{O}{\underset{\underset{OBr}{|}}{C}} + AgBr$

(b) $\phi-CH_2\overset{O}{\underset{\underset{OBr}{|}}{C}} \longrightarrow \phi-CH_2\overset{O}{\underset{\underset{O\cdot}{|}}{C}} + Br\cdot$

(c) $\phi-CH_2\!\!-\!\!\overset{O}{\underset{\underset{O\cdot}{|}}{C}} \longrightarrow \phi-CH_2\cdot + CO_2$

(d) $\phi-CH_2\cdot + \phi-CH_2\overset{O}{\underset{\underset{OBr}{|}}{C}} \longrightarrow \phi-CH_2Br + \phi-CH_2\overset{O}{\underset{\underset{O\cdot}{|}}{C}}$

(e) Repeat Steps (c) and (d).

This is an example of the Hunsdiecker reaction.

141. The label appears in the ester.

$$\phi-\overset{O}{\underset{\|}{C}}-OH \underset{}{\overset{H^\oplus}{\rightleftharpoons}} \left[\phi-\overset{\overset{\oplus}{O}H}{\underset{\|}{C}}-OH \longleftrightarrow \phi-\overset{OH}{\underset{\underset{OH}{|}}{\overset{|}{C}\oplus}} \right]$$

$$\updownarrow CH_3CH_2-{}^{18}\underline{O}-H$$

$$\phi-\overset{|\overline{O}H}{\underset{\underset{\oplus OH_2}{|}}{C}}-{}^{18}O-CH_2CH_3 \underset{}{\overset{H^\oplus \text{ transfer}}{\rightleftharpoons}} \phi-\overset{OH}{\underset{\underset{|OH}{|}}{C}}-\overset{\oplus}{{}^{18}O}-CH_2CH_3$$

$$\updownarrow$$

$$\phi-\overset{\overset{\oplus}{O}H}{\underset{\|}{C}}-{}^{18}OCH_2CH_3 \underset{}{\overset{-H^\oplus}{\rightleftharpoons}} \phi-\overset{O}{\underset{\|}{C}}-{}^{18}OCH_2CH_3$$

142.

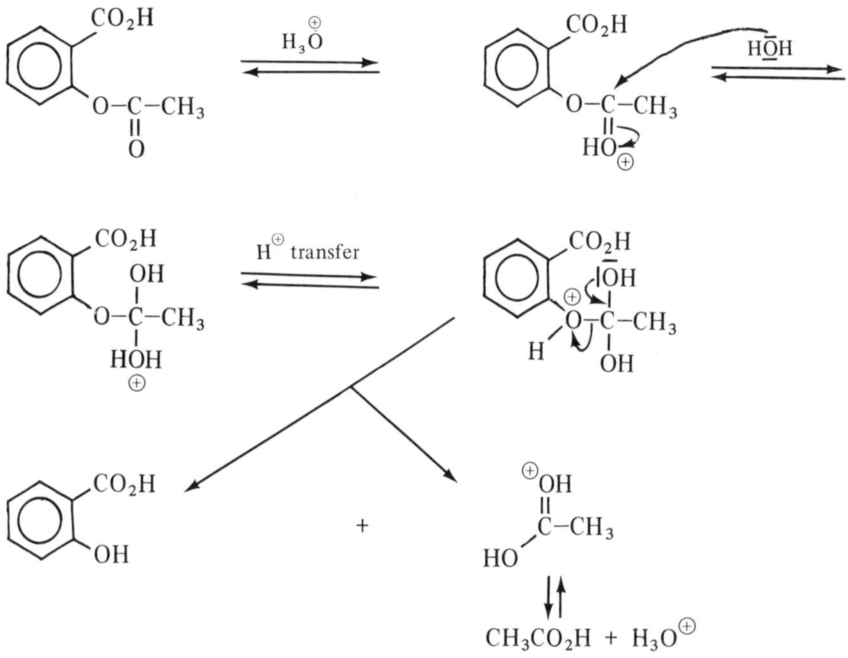

$$\text{2-(acetyloxy)benzoic acid} \xrightleftharpoons{H_3O^{\oplus}} \text{protonated ester} \xrightleftharpoons{H\underline{O}H} $$

tetrahedral intermediate $\xrightleftharpoons{H^{\oplus} \text{ transfer}}$ protonated tetrahedral intermediate

$\longrightarrow$ salicylic acid (2-hydroxybenzoic acid) + protonated acetic acid

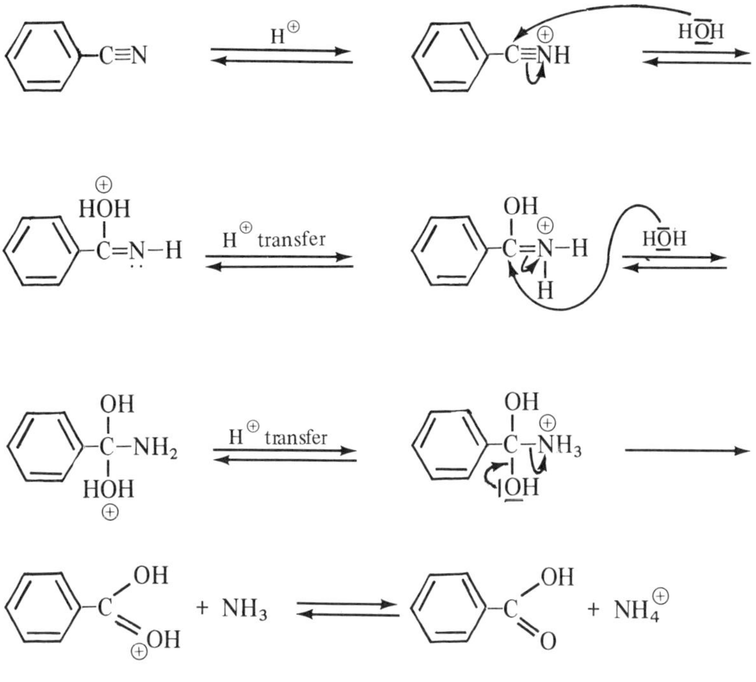

$$CH_3\overset{\overset{\oplus}{O}H}{\underset{HO}{C}}-CH_3 \rightleftharpoons CH_3CO_2H + H_3O^{\oplus}$$

143.

$$PhC\equiv N \xrightleftharpoons{H^{\oplus}} PhC\equiv \overset{\oplus}{N}H \xrightleftharpoons{H\underline{O}H}$$

$$Ph-\underset{\underset{..}{}}{C}(\overset{\oplus}{O}H_2)=N-H \xrightleftharpoons{H^{\oplus}\ \text{transfer}} Ph-\underset{H}{\overset{OH}{C}}=\overset{\oplus}{N}(H)-H \xrightleftharpoons{H\underline{O}H}$$

$$Ph-\underset{\overset{\oplus}{HOH}}{\overset{OH}{C}}-NH_2 \xrightleftharpoons{H^{\oplus}\ \text{transfer}} Ph-\underset{\underline{OH}}{\overset{OH}{C}}-\overset{\oplus}{N}H_3 \longrightarrow$$

$$Ph-\underset{\underset{\oplus}{OH}}{\overset{OH}{C}} + NH_3 \rightleftharpoons Ph-\underset{O}{\overset{OH}{C}} + NH_4^{\oplus}$$

162 SOLUTIONS WITH STRUCTURAL FORMULAS

144. A > C > B

Compound A is slightly more acidic than compound C, but C is more acidic than compound B. The resonance effect of the —OH group opposes the inductive effect. The para —OH group in compound B stabilizes the acid (more than the corresponding carboxylate anion) thereby making the acidic hydrogen in compound B less labile than in compound C. In compound A a similar resonance effect, however, cannot be transmitted to the carboxylic acid group from the meta position, thus leaving only the effect of the hydroxyl determining the acidity of compound A and the increased acidity of A relative to compound C. In compound A, where the resonance effect of the —OH cannot release electrons to the carboxylic acid moiety, the inductive effect is more important.

145. Start with:

ϕ—CO₂H

ϕ—OH

ϕ—ϕ

in ether

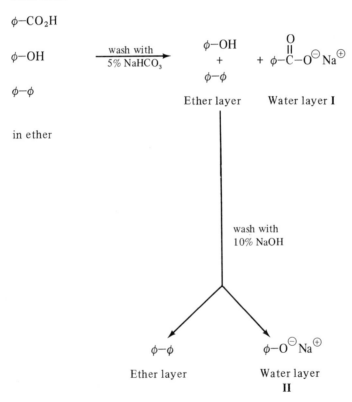

Acidifying aqueous layers **I** and **II** produces benzoic acid and phenol, respectively.

146. Internal hydrogen bonding can occur in the conjugate base of phthalic acid but not in the conjugate base of terephthalic acid. The separation of the carboxylic acid groups in terephthalic acid is too great for intramolecular hydrogen bonding to occur.

Conjugate base of phthalic acid

On the other hand, loss of the second proton from phthalic acid is more difficult than the loss of the second proton from terephthalic acid because the resulting dianion from phthalic acid has negative charges that are too close to each other.

(from phthalic acid)   (from terephthalic acid)

147.

164 SOLUTIONS WITH STRUCTURAL FORMULAS

148. B > A > C

149.

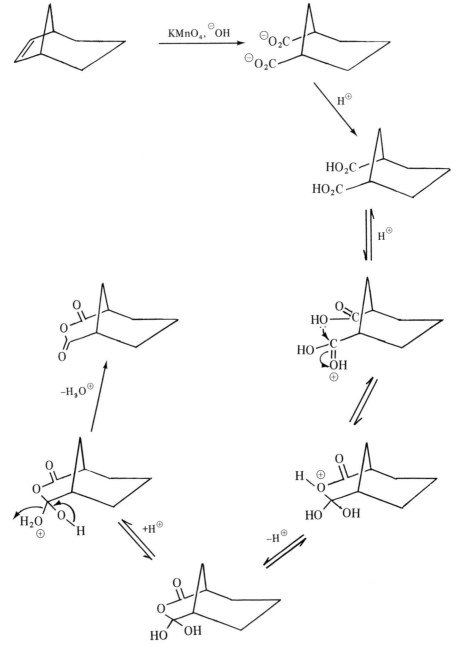

150.

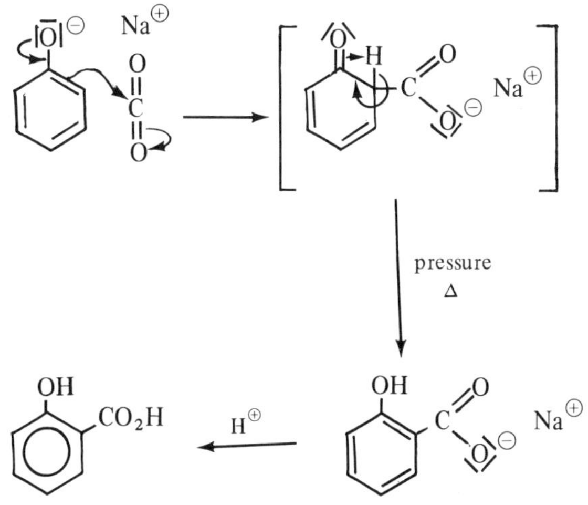
pressure
Δ

Salicylic acid

151.

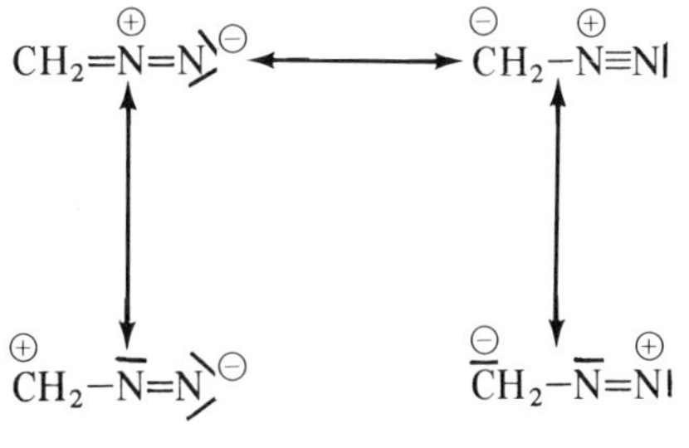

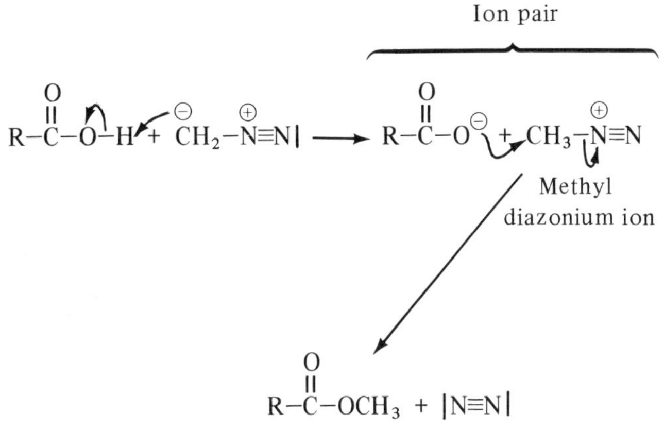
Ion pair
Methyl
diazonium ion

152.

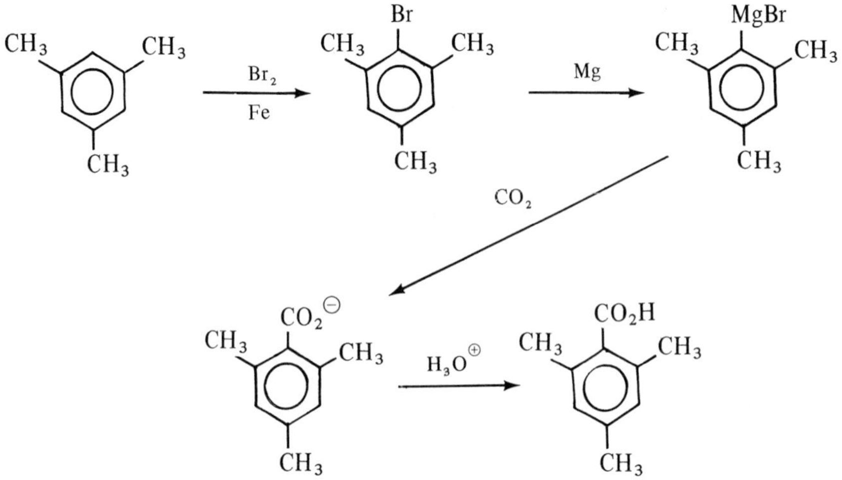

Mesitylene →(Br₂/Fe)→ 2-bromomesitylene →(Mg)→ mesitylmagnesium bromide →(CO₂)→ mesitylcarboxylate anion →(H₃O⁺)→ mesitoic acid (2,4,6-trimethylbenzoic acid)

153.

Phthalic anhydride + φ–H →(AlCl₃)→ **A** (2-benzoylbenzoic acid)

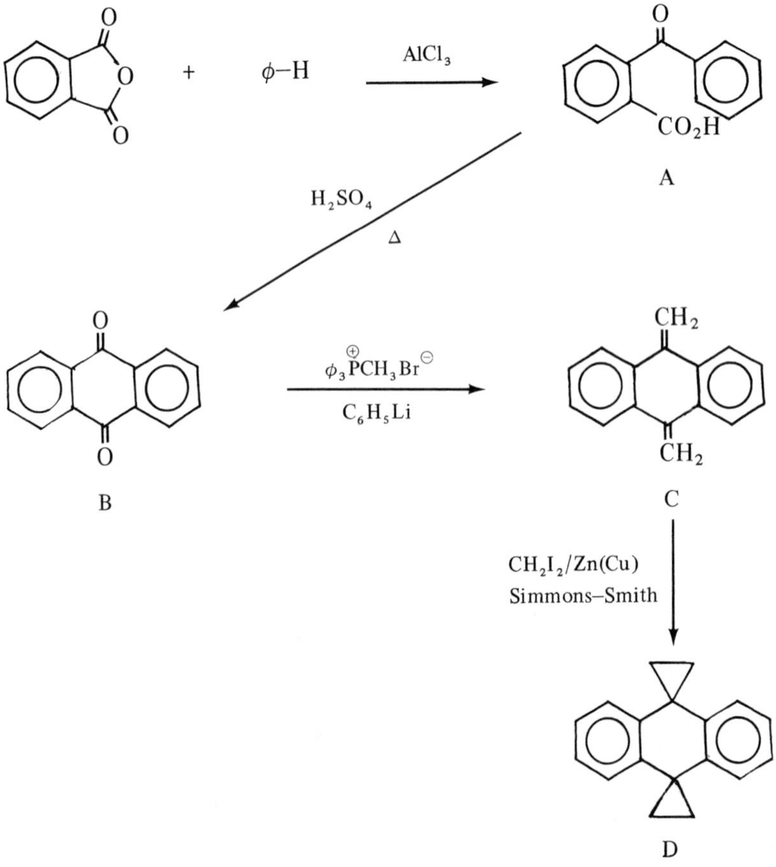

A →(H₂SO₄, Δ)→ **B** (anthraquinone)

B →(φ₃P⁺CH₃ Br⁻ / C₆H₅Li)→ **C** (9,10-dimethyleneanthracene)

C →(CH₂I₂/Zn(Cu), Simmons–Smith)→ **D** (dispiro biscyclopropane product)

154.

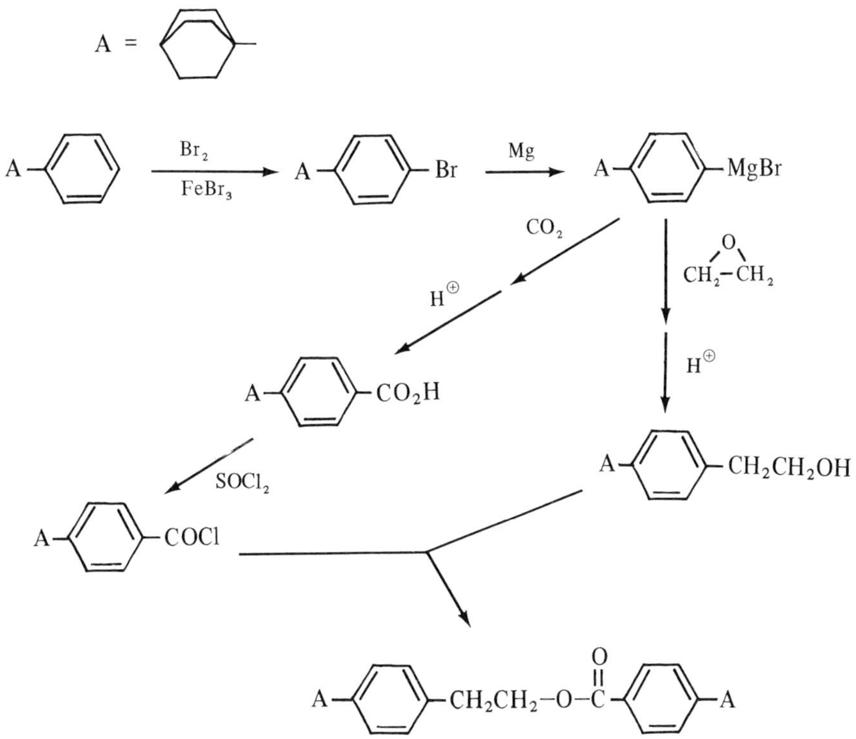
A =
A
Br2
FeBr3
A
Br
Mg
A
MgBr
CO2
H⊕
CH2–CH2
O
H⊕
A
CO2H
SOCl2
A
CH2CH2OH
A
COCl
A
CH2CH2–O–C
O
A

155.

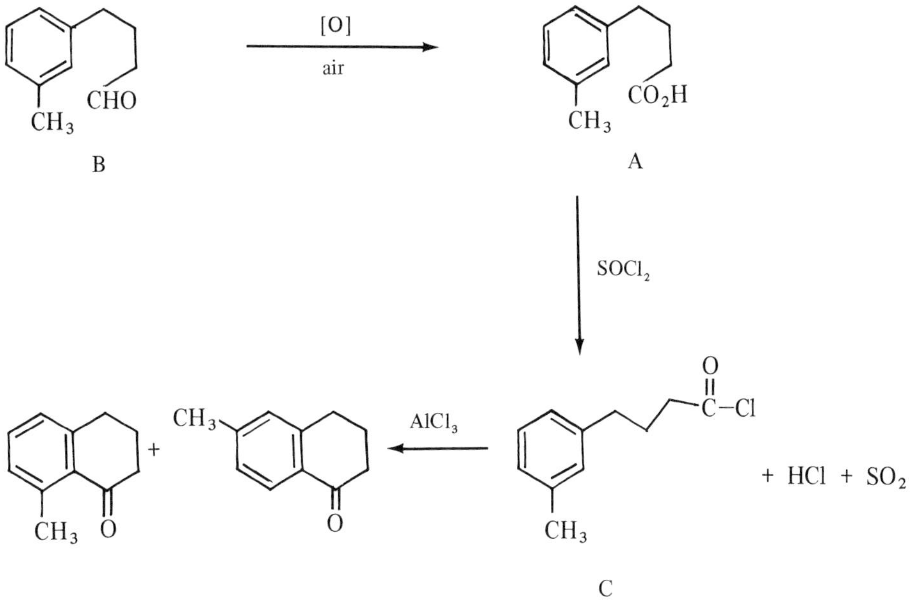
CHO
CH3
B
[O]
air
CO2H
CH3
A
SOCl2
O
C–Cl
CH3
C
+ HCl + SO2
AlCl3
CH3
O
+
CH3
O

156.

157.

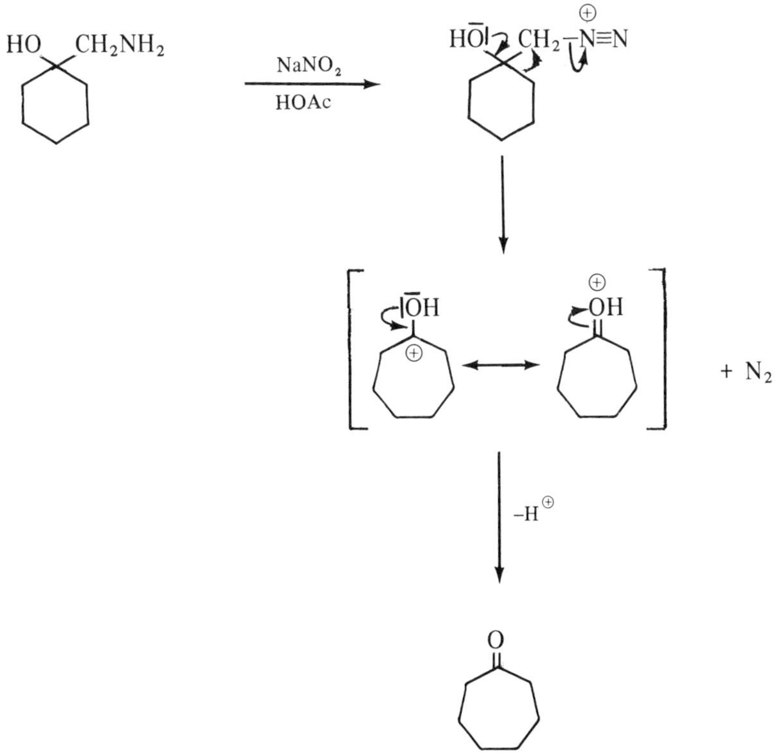

[Reaction scheme: 1-(aminomethyl)cyclohexanol treated with NaNO₂/HOAc forms a diazonium intermediate, which through semipinacol-type ring expansion with loss of N₂ gives a resonance-stabilized oxocarbenium ion (cycloheptane ring with +OH), then loss of H⁺ gives cycloheptanone.]

+ N₂

−H⁺

158. CH₃−N=C=O

Methyl isocyanate

This compound caused one of the worst chemical disasters in history in Bhopal, India.

159.

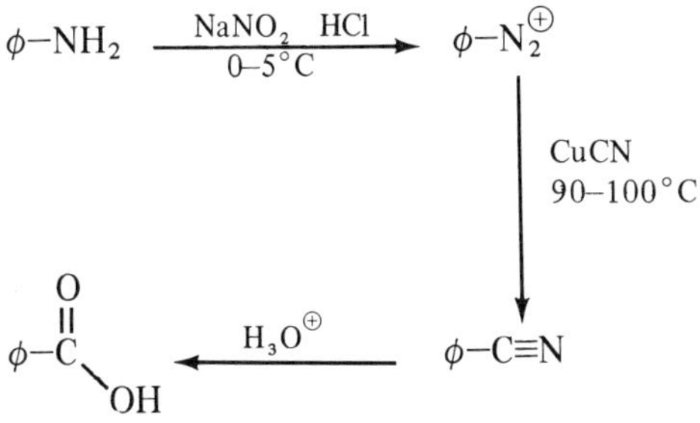

$\phi-NH_2 \xrightarrow[0-5°C]{NaNO_2 \ HCl} \phi-N_2^{\oplus} \xrightarrow[90-100°C]{CuCN} \phi-C\equiv N \xrightarrow{H_3O^{\oplus}} \phi-\underset{OH}{\overset{O}{\underset{\|}{C}}}$

160.

161.

SOLUTIONS WITH STRUCTURAL FORMULAS

162.

163.

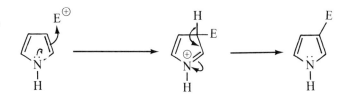

164. & **165.**

166.

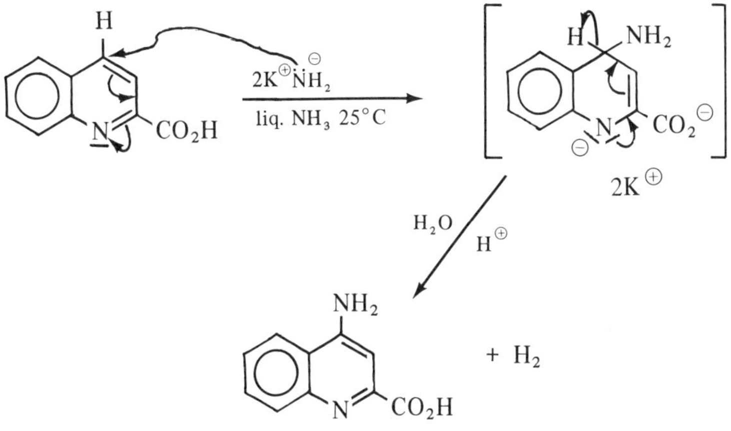

[Quinoline-2-carboxylic acid + 2K⁺⁻NH₂, liq. NH₃, 25°C → intermediate with NH₂ added at 4-position, N⁻, CO₂⁻, 2K⁺ → H₂O/H⁺ → 4-amino-quinoline-2-carboxylic acid + H₂]

167.

A: Ph–C(=O)–Ph ⟶ B: Ph–C(=N–OH)–Ph ⟶

C: Ph–C(=O)–N(H)–Ph ⟶ D: Ph–NH₂ ⟶

E: Ph–N(H)–C(=O)–CH₃

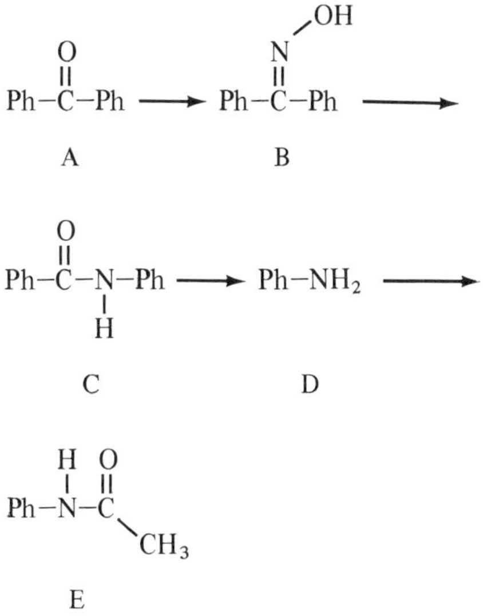

Note: Ph = φ– = C₆H₅–

168.

SOLUTIONS WITH STRUCTURAL FORMULAS

169. [Reaction mechanism showing phenyllithium addition to pyridine at the 2-position, proceeding through three resonance structures of the anionic intermediate, followed by loss of hydride to give 2-phenylpyridine + LiH]

170. [Hofmann rearrangement mechanism of hexanamide:]

- Hexanamide + Br$_2$/NaOH → N-bromoamide + H$_2$O + NaBr
- Deprotonation by $^{\ominus}$OH
- Loss of Br$^{\ominus}$ gives a nitrene
- Rearrangement to alkyl isocyanate (N=C=O)
- Addition of H$_2$O (with $^{\ominus}$OH) gives carbamic acid
- $-H^{\oplus}$ gives carbamate
- $^{\ominus}$OH addition to C=O
- H$^{\oplus}$ transfer gives pentylamine (R–NH$_2$) + CO$_3^{2\ominus}$

176 SOLUTIONS WITH STRUCTURAL FORMULAS

171.

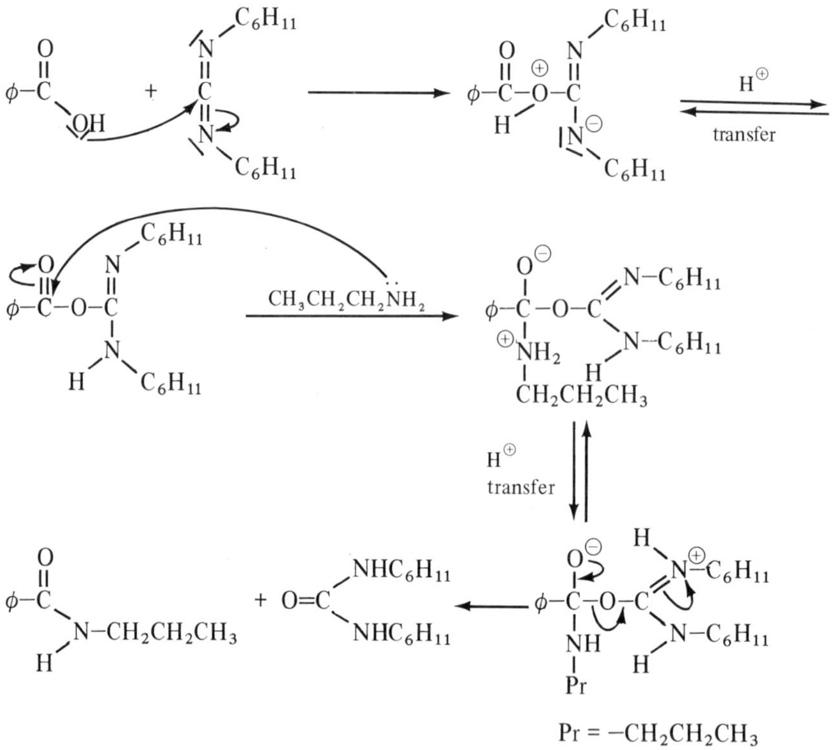

$$\phi-\underset{OH}{\overset{O}{\underset{\|}{C}}} + \underset{C_6H_{11}}{\overset{C_6H_{11}}{\underset{\|}{N}}}\overset{}{\underset{\|}{C}} \longrightarrow \phi-\underset{H}{\overset{O}{\underset{\|}{C}}}-O-\underset{\underset{C_6H_{11}}{\|}}{\overset{\overset{N-C_6H_{11}}{\|}}{C}} \underset{\text{transfer}}{\overset{H^{\oplus}}{\rightleftharpoons}}$$

$$\phi-\underset{\underset{\underset{C_6H_{11}}{|}}{N}}{\overset{O}{\underset{\|}{C}}}-O-\underset{H}{\overset{N-C_6H_{11}}{\underset{\|}{C}}} \xrightarrow{CH_3CH_2CH_2NH_2} \phi-\overset{O^{\ominus}}{\underset{\underset{\underset{CH_2CH_2CH_3}{|}}{\overset{\oplus}{NH_2}}}{C}}-O-\underset{H}{\overset{N-C_6H_{11}}{\underset{\|}{C}}}$$

$$\underset{\text{H}^{\oplus} \text{ transfer}}{\updownarrow}$$

$$\phi-\underset{H}{\overset{O}{\underset{\|}{C}}}-\underset{CH_2CH_3}{\overset{}{\underset{\|}{N}}} + O=C\underset{NHC_6H_{11}}{\overset{NHC_6H_{11}}{\underset{}{\diagup}}} \longleftarrow \phi-\overset{O^{\ominus}}{\underset{\underset{Pr}{\overset{|}{NH}}}{C}}-O-\underset{H}{\overset{\overset{H}{\underset{\|}{N}}\overset{\oplus}{-}C_6H_{11}}{\underset{\|}{C}}}$$

$$Pr = -CH_2CH_2CH_3$$

172.

[Mechanism showing Pictet-Spengler type cyclization: dimethoxyphenethylamine + aldehyde with dimethoxybenzyl group → iminium ion → cyclized tetrahydroisoquinoline after −H⁺]

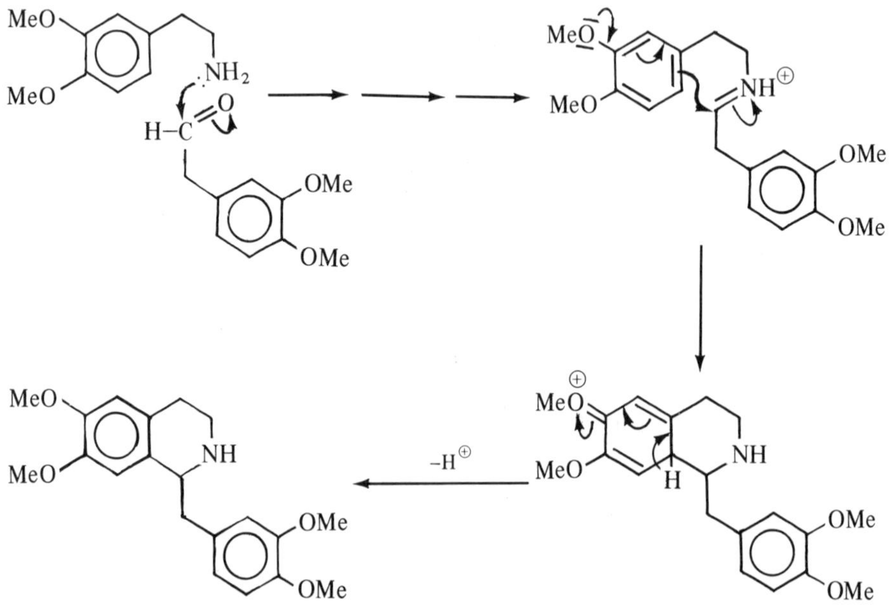

SOLUTIONS WITH STRUCTURAL FORMULAS 177

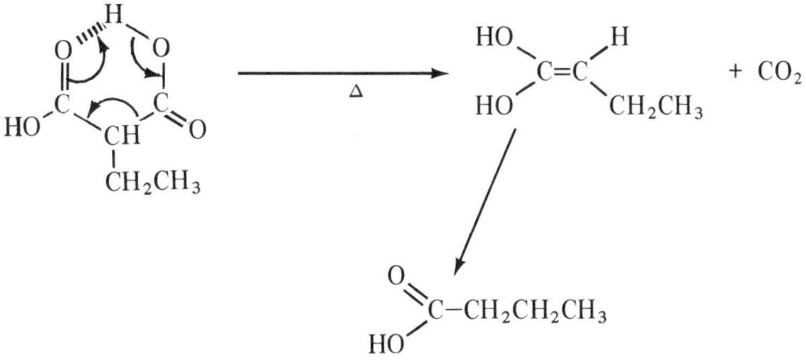

173.

$$\underset{\underset{CH_2CH_3}{|}}{\underset{HO}{\overset{O}{\|}}{C}}-\underset{\underset{}{CH}}{\overset{\overset{H}{\overset{|}{O}}\cdots\overset{O}{}}{}}-\overset{O}{\underset{}{\|}}{C}=O \quad \xrightarrow{\Delta} \quad \underset{HO}{\overset{HO}{}}C=\underset{CH_2CH_3}{\overset{H}{}} \quad + \quad CO_2$$

$$\downarrow$$

$$\underset{HO}{\overset{O}{\|}}{C}-CH_2CH_3$$

174.

$$EtO-\overset{O}{\underset{\|}{C}}-CH_2-\overset{O}{\underset{\|}{C}}-OEt \quad \underset{EtOH}{\overset{Et_2NH}{\rightleftharpoons}} \quad EtO-\overset{O}{\underset{\|}{C}}-\overset{\ominus}{\underset{..}{CH}}-\overset{O}{\underset{\|}{C}}-OEt \quad + \quad Et_2\overset{\oplus}{N}H_2$$

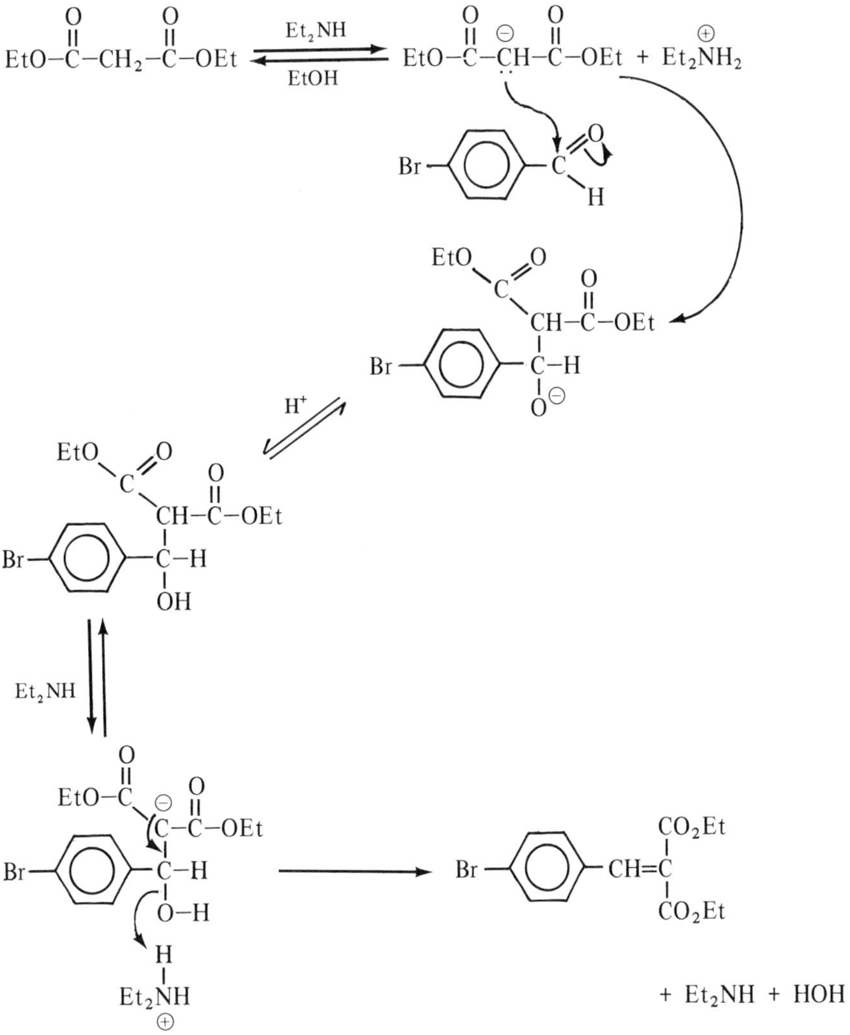

Br—C$_6$H$_4$—CHO (attacked by carbanion)

$$\underset{Br—C_6H_4—}{\overset{EtO}{\underset{|}{C}}=O} \overset{O}{\underset{\|}{C}}-OEt \quad\text{with}\; \underset{O^\ominus}{\overset{|}{CH}}$$

$$\updownarrow H^+$$

EtO—C(=O)—CH(—C(=O)—OEt)—CH(—C$_6$H$_4$Br)—OH

$$\updownarrow Et_2NH$$

EtO—C(=O)—$\overset{\ominus}{C}$(—C(=O)—OEt)—CH(—C$_6$H$_4$Br)—O—H ··· $\overset{\oplus}{N}$HEt$_2$

$$\longrightarrow \quad Br—C_6H_4—CH=C\underset{CO_2Et}{\overset{CO_2Et}{<}} \quad + \quad Et_2NH \; + \; HOH$$

175.

176.

177.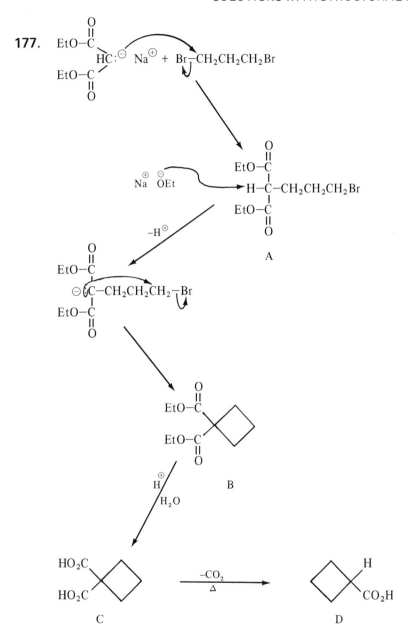

180 SOLUTIONS WITH STRUCTURAL FORMULAS

178.

$$CH_3-\overset{O}{\overset{\|}{C}}-CH_2-\overset{O}{\overset{\|}{C}}-OEt \xrightarrow{NaOEt} CH_3-\overset{O}{\overset{\|}{C}}-\overset{\ominus}{\underset{..}{C}H}-\overset{O}{\overset{\|}{C}}-OEt$$

$$\xrightarrow{Br-CH_2-\overset{O}{\overset{\|}{C}}-CH_3}$$

A:
$$CH_3-\overset{O}{\overset{\|}{C}}-\underset{\underset{\underset{CH_3}{\mid}}{\underset{C=O}{\mid}}}{\underset{CH_2}{\overset{\mid}{CH}}}-\overset{O}{\overset{\|}{C}}-OEt$$

$\xrightarrow{\text{1. dil. NaOH} \atop \text{2. } H_3O^{\oplus}}$

B:
$$CH_3-\overset{O}{\overset{\|}{C}}-\underset{\underset{\underset{CH_3}{\mid}}{\underset{C=O}{\mid}}}{\underset{CH_2}{\overset{\mid}{CH}}}-\overset{O}{\overset{\|}{C}}-OH$$

$\xrightarrow{-CO_2 \atop \Delta}$

C:
$$CH_3-\overset{O}{\overset{\|}{C}}-CH_2CH_2-\overset{O}{\overset{\|}{C}}-CH_3$$

179. A and B are reducing sugars.

B: pyranose form (HOCH₂, OH, H on ring) ⇌ A: open chain CHO, H—OH, HO—H, H—OH, H—OH, CH₂OH $\xrightarrow[\text{NaOH}]{Cu^{2+}}$ CO₂⁻, H—OH, HO—H, H—OH, H—OH, CH₂OH + Cu₂O

180.

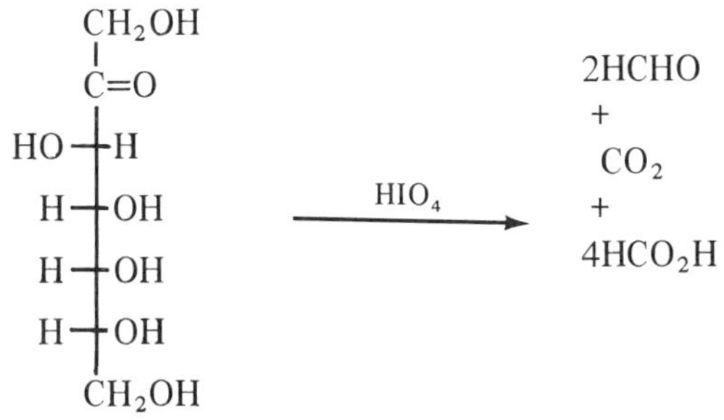

$$\begin{array}{c} CH_2OH \\ | \\ C=O \\ HO-H \\ H-OH \\ H-OH \\ H-OH \\ | \\ CH_2OH \end{array} \xrightarrow{HIO_4} \begin{array}{c} 2HCHO \\ + \\ CO_2 \\ + \\ 4HCO_2H \end{array}$$

181. Hydrolysis of the sucrose solution is accelerated by heat and dilute acid. The vinegar supplies the dilute acid. Hydrolysis of 1 mole of sucrose (sweetness factor 1.45) produces 1 mole of glucose and 1 mole of fructose (total sweetness factor 2.65). The *cooked* fudge tastes sweeter than the combination of the uncooked ingredients. There is a greater number of moles of sugar present in the fudge than in the starting material.

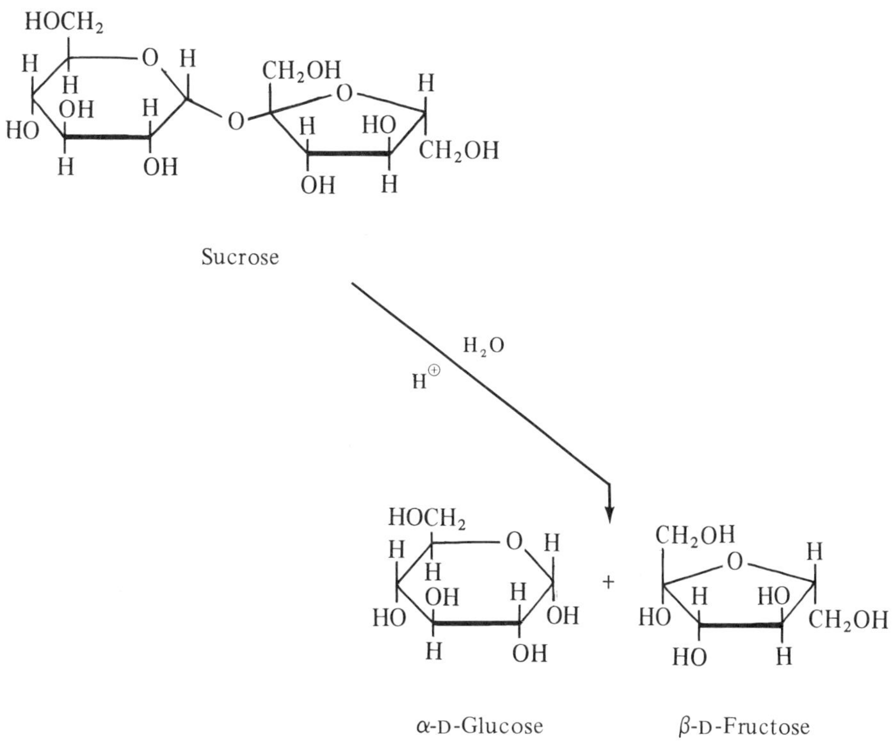

Sucrose

$\xrightarrow{H_2O, H^{\oplus}}$

α-D-Glucose + β-D-Fructose

182 SOLUTIONS WITH STRUCTURAL FORMULAS

182.

$$\begin{array}{c} \text{CHO} \\ \text{H–C–OH} \\ \text{HO–C–H} \\ \text{H–C–OH} \\ \text{H–C–OH} \\ \text{CH}_2\text{OH} \end{array} \quad \xrightleftharpoons{\text{NaOH}} \quad \begin{array}{c} \text{CHO} \\ \text{H–C–OH} \\ \text{HO–C–H} \\ \text{H–C–O}^{\ominus} \\ \text{H–C–OH} \\ \text{CH}_2\text{OH} \end{array} + \text{H}_2\text{O} + \text{Na}^{\oplus}$$

splits into:

$$\begin{array}{c} \text{CHO} \\ \text{H–C–OH} \\ \text{HO–C–H} \\ ^{\ominus} \end{array} \quad \rightleftharpoons \quad \begin{array}{c} \text{H}\diagdown\text{C}{=}\text{O} \\ \text{H–C–OH} \\ \text{CH}_2\text{OH} \end{array} \quad \text{A}$$

$\text{H}^{\oplus}$ loss

$$\begin{array}{c} \text{H}\diagdown\text{C–O}^{\ominus} \\ \parallel \\ \text{C–OH} \\ \text{CH}_2\text{OH} \end{array} \longleftrightarrow \begin{array}{c} \text{H}\diagdown\text{C}{=}\text{O} \\ ^{\ominus}\text{C–OH} \\ \text{CH}_2\text{OH} \end{array}$$

proton transfer $\updownarrow$

$$\begin{array}{c} \text{H}\diagdown\text{C}\diagup\text{OH} \\ \parallel \\ \text{C–O}^{\ominus} \\ \text{CH}_2\text{OH} \end{array} \longleftrightarrow \begin{array}{c} \text{H}\diagdown\text{C}\diagup\text{OH} \\ ^{\ominus}\text{C} \\ \text{C}{=}\text{O} \\ \text{CH}_2\text{OH} \end{array} \quad \text{B}$$

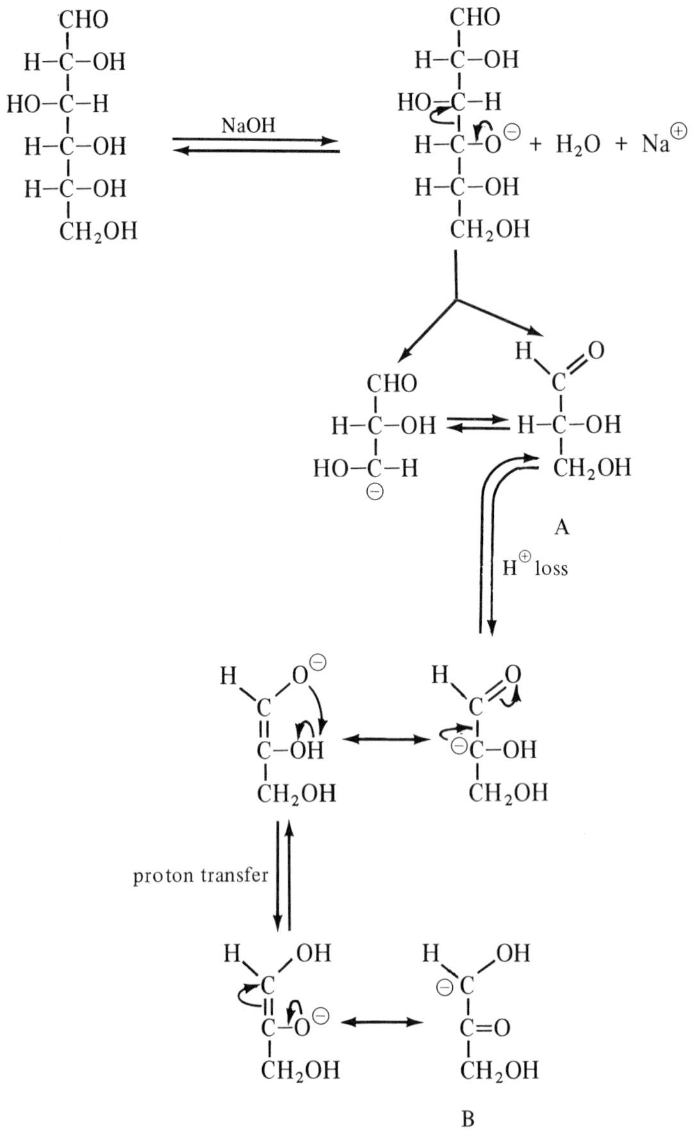

A + B $\longrightarrow$ D-fructose (via aldol condensation)

183.

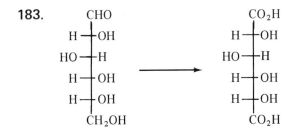

D-Glucose → D-Glucaric acid

L-Gulose → D-Glucaric acid ≡ D-Glucaric acid

184.

$$\begin{array}{l} CH_2-O-\overset{O}{\underset{\|}{C}}-R_1 \\ | \\ CH-O-\overset{O}{\underset{\|}{C}}-R_2 \\ | \\ CH_2-O-\overset{O}{\underset{\|}{C}}-R_3 \end{array} \quad \begin{array}{l} CH_2-O-\overset{O}{\underset{\|}{C}}-R_2 \\ | \\ CH-O-\overset{O}{\underset{\|}{C}}-R_1 \\ | \\ CH_2-O-\overset{O}{\underset{\|}{C}}-R_3 \end{array} \quad \begin{array}{l} CH_2-O-\overset{O}{\underset{\|}{C}}-R_1 \\ | \\ CH-O-\overset{O}{\underset{\|}{C}}-R_3 \\ | \\ CH_2-O-\overset{O}{\underset{\|}{C}}-R_2 \end{array}$$

185. Compound **II**, which has two lysine residues and is therefore the most basic of the three peptides, migrates most rapidly. At pH 1.5 three formal positive charges exist on **II**. (The first is on the α-amino group of valine. The next two are on the ε-amino groups of the lysine residues.)

184 SOLUTIONS WITH STRUCTURAL FORMULAS

186. Any amino group is free to react with DNFB by displacing fluorine via nucleophilic aromatic substitution. Ornithine has an γ-amino group which, in addition to the α-amino group of valine, also can react with DNFB. Following acid hydrolysis of the peptide, the two derivatized products that result are

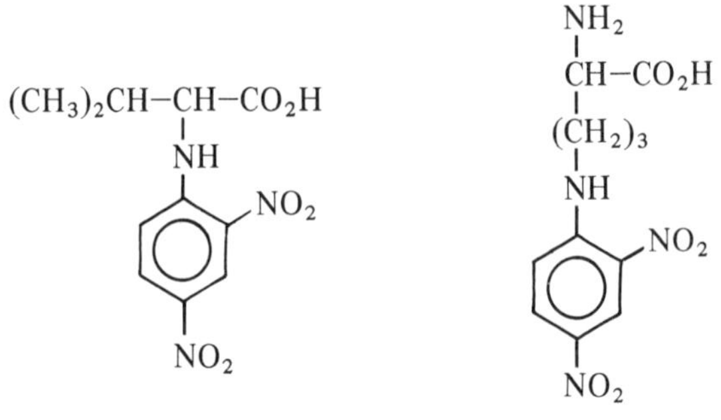

$(CH_3)_2CH-CH-CO_2H$
      |
      NH
      |
   (2,4-dinitrophenyl)

DNP–Val

NH_2
 |
$CH-CO_2H$
 |
$(CH_2)_3$
 |
 NH
 |
(2,4-dinitrophenyl)

γ-DNP–Orn

187.

[Mechanism scheme for CNBr cleavage of methionine-containing peptide: starting peptide with $-S-CH_3$ side chain reacts with Br–CN to give sulfonium intermediate $N\equiv C-S^+(CH_3)-CH_2-$; intramolecular displacement by carbonyl oxygen with loss of $N\equiv C-S-CH_3$ gives iminolactone; addition of H_2O, followed by H^+ transfer, yields tetrahedral intermediate; collapse with loss of NH_2R gives the homoserine lactone product.]

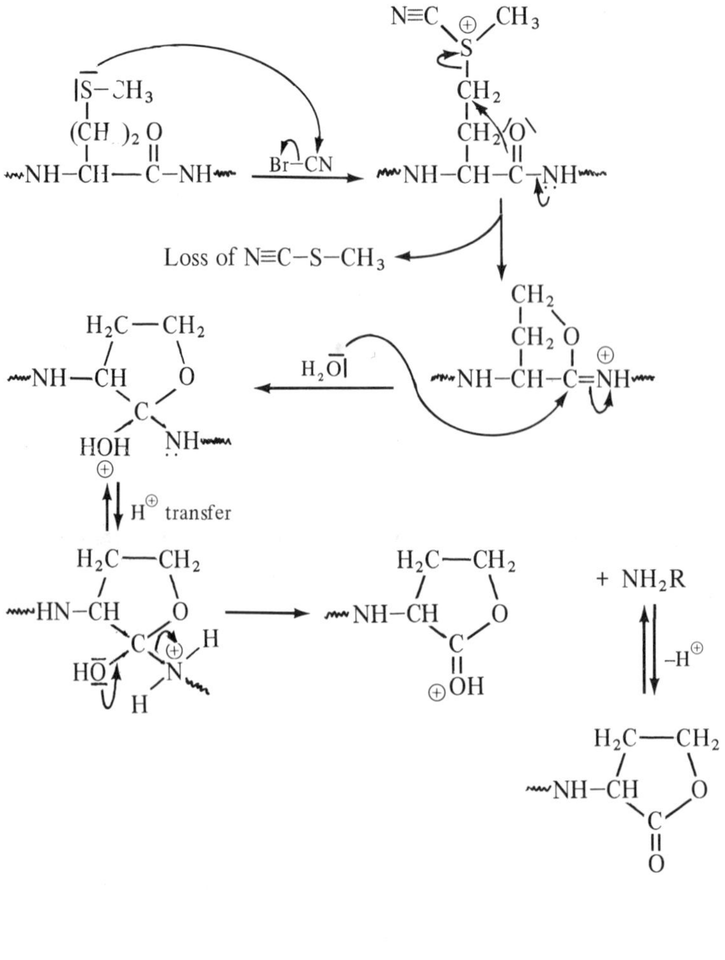

188.

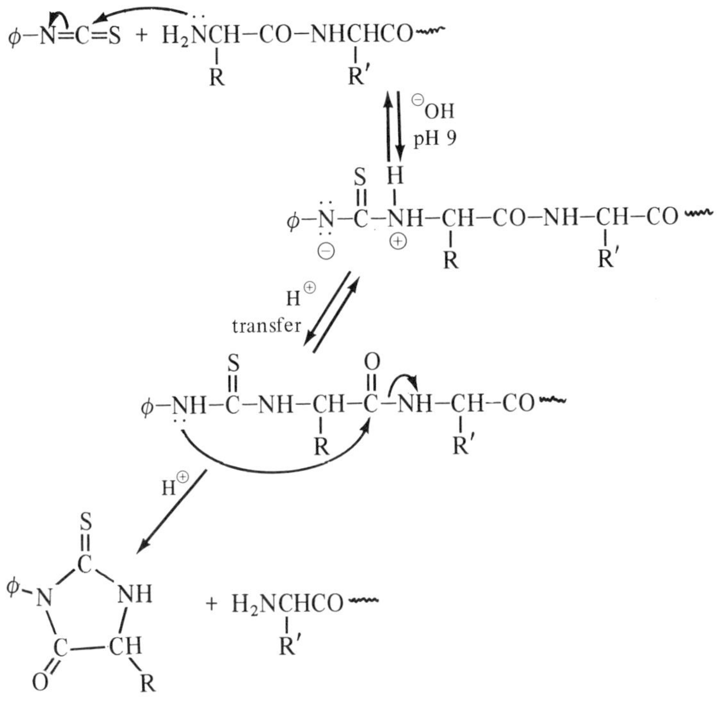

$$\phi-N=C=S + H_2NCH(R)-CO-NHCH(R')CO\text{---}$$

$\updownarrow\ ^{\ominus}OH,\ pH\ 9$

$$\phi-\overset{\ominus}{N}-\underset{\underset{S}{\|}}{C}-\overset{\oplus}{N}H-CH(R)-CO-NH-CH(R')-CO\text{---}$$

$\updownarrow\ H^{\oplus}\ \text{transfer}$

$$\phi-NH-\underset{\underset{S}{\|}}{C}-NH-CH(R)-\underset{\underset{O}{\|}}{C}-NH-CH(R')-CO\text{---}$$

$\downarrow H^{\oplus}$

Phenylthiohydantoin + $H_2NCH(R')CO\text{---}$

(cyclic phenylthiohydantoin structure with $\phi-N$, $C=S$, NH, $C=O$, $CH-R$)

189.

$$NH_2CH_2CO_2H + \phi-CH_2OCOCl \longrightarrow \phi-CH_2-O-\overset{\underset{\|}{O}}{C}-NHCH_2CO_2H$$
A

$\downarrow SOCl_2$

$$\phi-CH_2-O-\overset{\underset{\|}{O}}{C}-NHCH_2\overset{\underset{\|}{O}}{C}-Cl$$
B

$\downarrow$ alanine

$$\phi-CH_2-O-\overset{\underset{\|}{O}}{C}-NHCH_2\overset{\underset{\|}{O}}{C}-NH-CH(CH_3)CO_2H$$
C

$\downarrow H_2/Pd$

$$H_2NCH_2\overset{\underset{\|}{O}}{C}-NH-CH(CH_3)CO_2H + \text{toluene} + CO_2$$
D   E

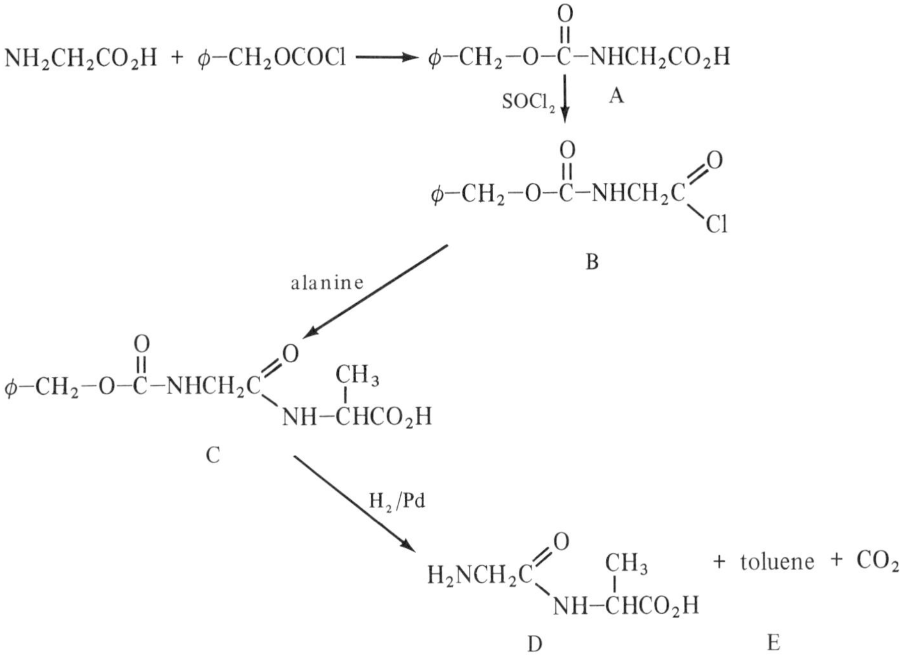

190.

191.

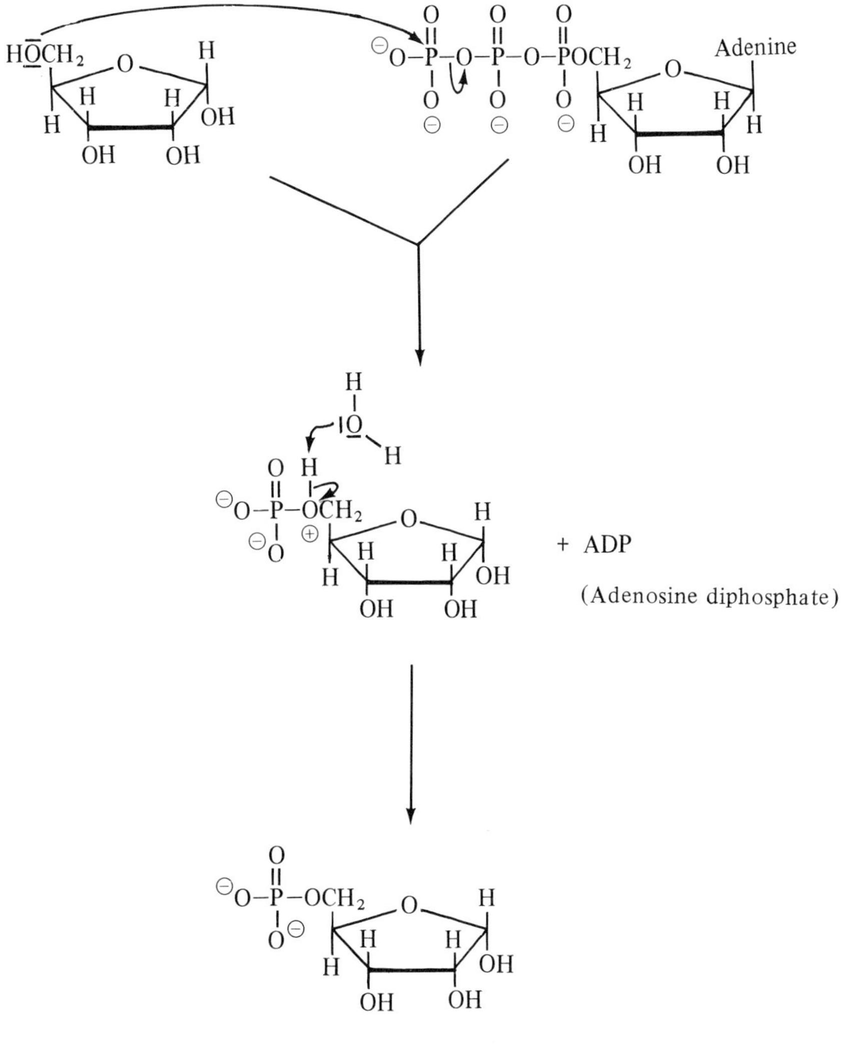

Ribose-5-phosphate

192.

193.

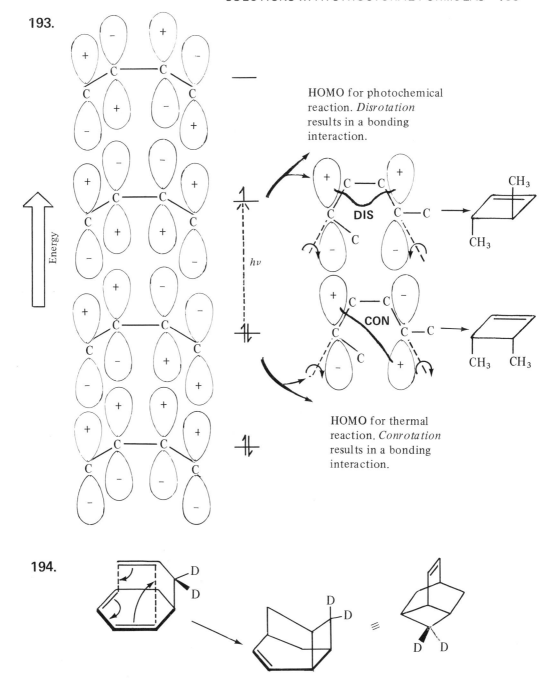

194.

195.

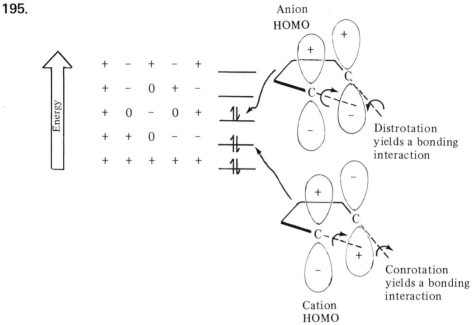

196.

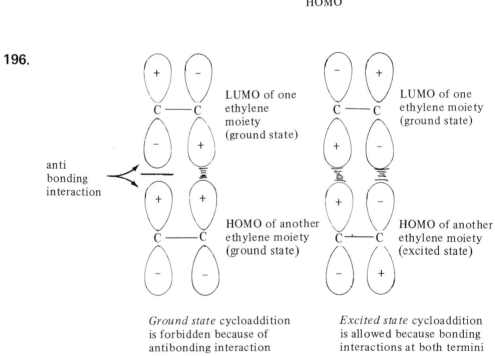

197.

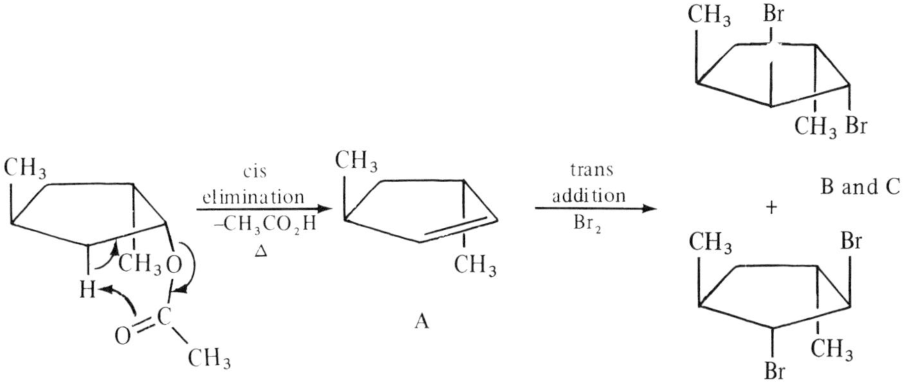
cis
elimination
$-CH_3CO_2H$
Δ
trans
addition
Br_2
A
B and C

198.

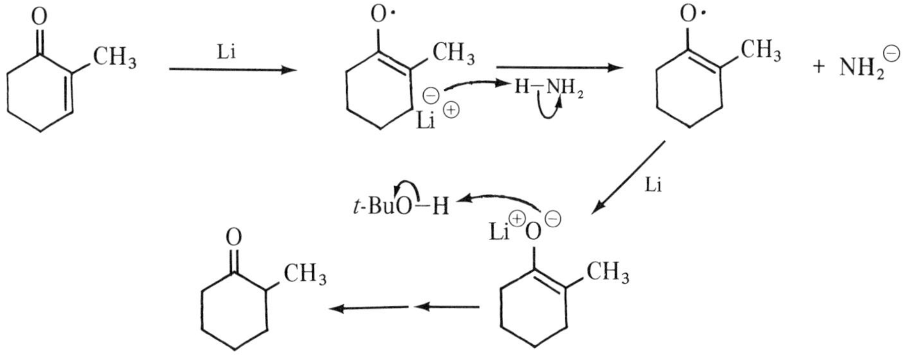
Li
Li
t-BuO–H

199.

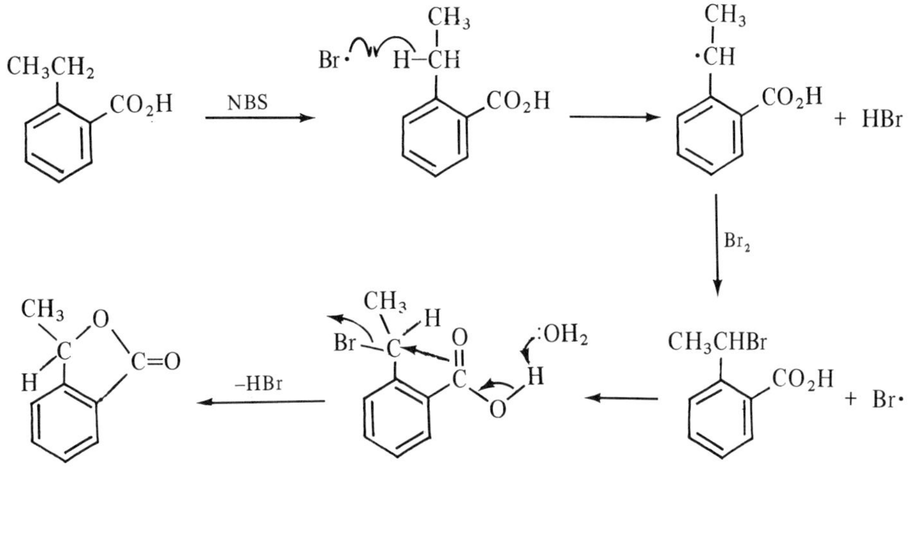
NBS
Br_2
–HBr

192 SOLUTIONS WITH STRUCTURAL FORMULAS

200.

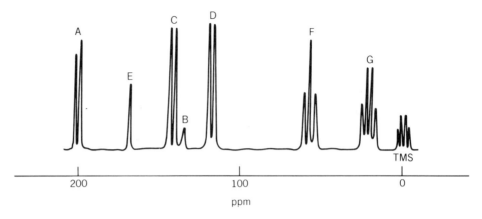

201.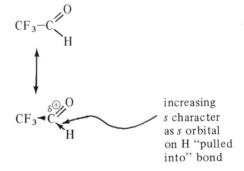

202. $^1J_{CH} \simeq 500\,X$
   $207 = 500\,X$
   $\therefore X = 0.4\,s$

increasing s character as s orbital on H "pulled into" bond

203.

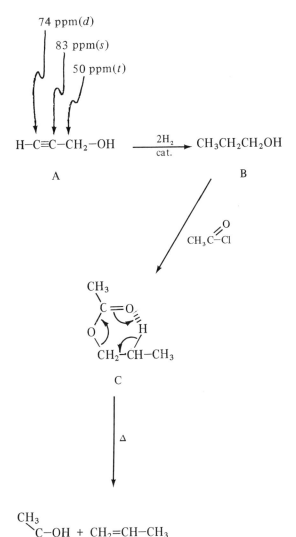

204. The data do not rule out

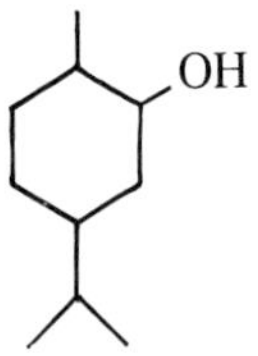
OH

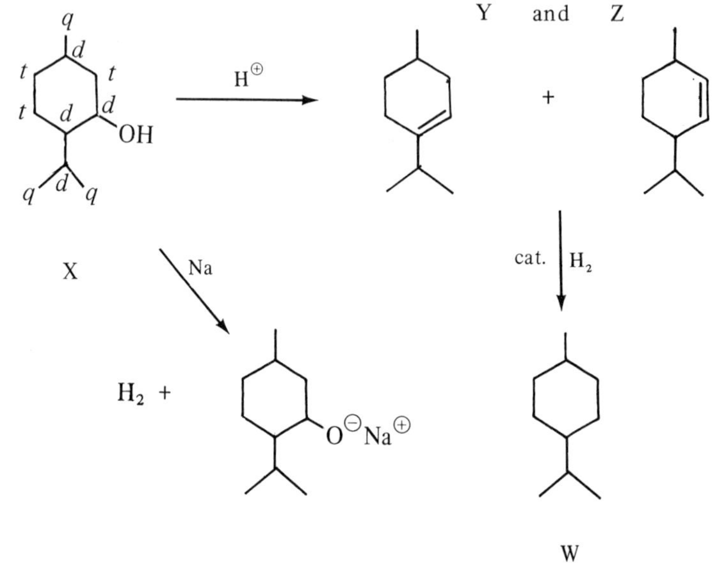
q
d
t
t
t
d
d
OH
q
d
q
X
$H^{\oplus}$
Y and Z
+
Na
cat.
H_2
H_2 +
$O^{\ominus}Na^{\oplus}$
W

205.

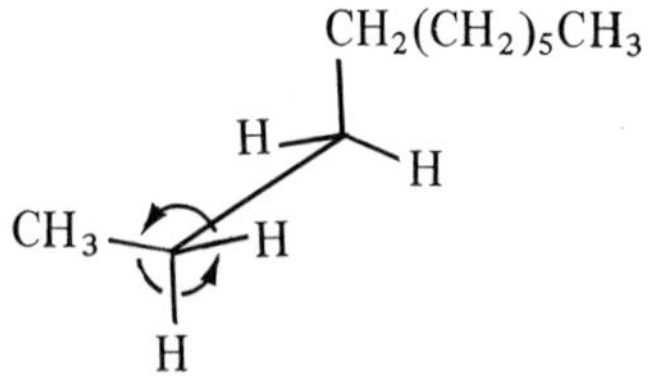
$CH_2(CH_2)_5CH_3$
H
H
CH_3
H
H

Rotation here is relatively easy

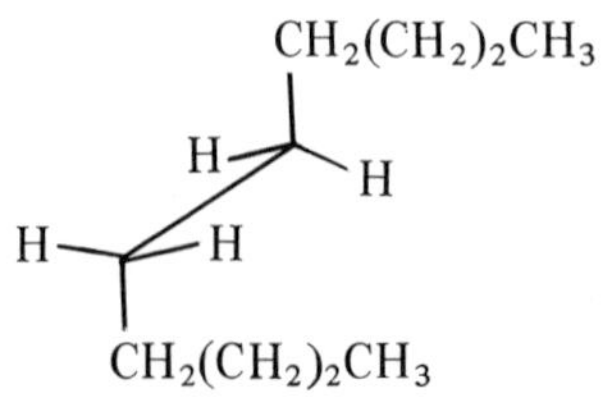
$CH_2(CH_2)_2CH_3$
H
H
H
H
$CH_2(CH_2)_2CH_3$

Motion more difficult—rotational barriers are higher. "Butyl group" mass requires more energy for rotation than "methyl group" to the left.

INDEX TO PROBLEMS

Numbers refer to problem numbers, not page numbers.

Acetal formation, mechanism, 115
Acetals, synthesis from ketones, 115
Acetate ion, 12
Acetate pyrolysis, mechanism and products, 197
Acetoacetic ester synthesis, 178
Acid chloride:
  hydrolysis, 136
  reactions, 137
Acid chlorides, reaction with:
  ammonia, 135
  ethanol, 135
  ethyl amine, 135
  sodium acetate, 135
Acidity:
  carboxylic acids, 144
  enols, 87
  phenols, 108
  pKa comparison, 9
  phthalic acid, 146
  substituent effects on, 144
  terephthalic acid, 146
Acids:
  decarboxylation, 173
  esterification with alcohols, 93
  synthesis, 93
Alcohol, synthesis, 45
Alcohols:
  esterification, 141
  nomenclature, 17
  reaction with chromic acid, 109
  reactions, 42, 91, 95, 102
  reactions with sulfuric acid, 107
  reaction with thionyl chloride, 96
  sulfonate ester formation, 102
  synthesis, 49, 90, 92
  synthesis using Grignard reagent, 97
Aldehydes:
  reaction with NaCN, 126
  reactions, 93, 129, 174
  reactions with Grignard reagent, 97
  reactivity, 127
  Tollens' reagent, 155
Aldehydes and Ketones:
  reactions, 124, 125, 128, 131, 132, 133
  synthesis, 128
Aldol condensation:
  of aldehydes and ketones, 124
  mechanism, 127
  retro, 128
Alkanes, 43
  bromination, 26
  bromination of methane, 23
  bromination of propane, 24
  chlorination, 22
  halogenation, 25
  nomenclature, 18
  reactions, 66
  synthesis, 116
Alkane synthesis, 15
Alkenes:
  bromination, 52, 197
  carbene addition, 67
  chlorohydrin formation, 59
  dimerization, 47
  diol synthesis, 54
  epoxide formation, 106
  polymerization, 63, 64
  reaction, 149
  reaction with HBr, 56
  reaction with ozone, 51, 55
  reactions, 45, 46, 48, 49, 50, 53, 57, 62, 65, 67
  reductions via lithium in ammonia, 198
  stereospecific synthesis, 70
  structure identification, 57
  synthesis, 67, 73, 107
Alkoxymercuration-demercuration, 48
Alkyl halides, reactions, 88, 98
Alkynes:
  addition of water, 69
  hydroboration, 68
  reaction with HCl, 71
  reactions, 68, 70, 72, 73
  synthesis, 72
Amide, hydrolysis, 136
Amide formation, mechanism, 171
Amides:
  reactions, 170
  synthesis, 147, 171
Amines:
  diazonium salt formation, 165
  Hinsberg test for amines, 162
  Hoffmann elimination, 164
  reaction with phosgene, 158
  reactions, 88, 137, 157, 159, 167, 168, 169, 171
  reactivity of pyrrole, 163
  relative reactivity with benzene-sulfonyl chloride, 163
  synthesis, 170
Amino acids:
  ninhydrin reaction, 190
  synthesis, 134
Anhydride, hydrolysis, 136
Anhydrides:
  reactions, 168
  synthesis, 149
Anisole, synthesis, 103

195

INDEX TO PROBLEMS

Aromatic, compounds:
   reactions, 80
   synthesis, 79, 80
Aromatic compounds:
   bromination, 83, 84
   nitration, 83
   reactions, 77, 79, 81, 82, 83, 85
   spectrometic identification, 200
   synthesis, 77, 83, 86
   synthesis of 1,3,5-trinitrobenzene, 86
Aromaticity, 76
ATP, 191

Baeyer-Villiger reaction, 121
Beckmann rearrangement, 167
   mechanism, 147
Benzoic acid synthesis from aniline, 159
Benzoin condensation, mechanism, 126
Bergmann synthesis, 189
Bicyclic compounds, nomenclature, 16
Biphenyl, solubility in ether, 145
Boiling points, structural effects on, 7
Bromination, carboxylic acids, 134

Cannizzaro reaction, mechanism, 129
Carbanion stability, 94
Carbaryl:
   mechanism of formation, 160
   synthesis, 137
   synthesis from methyl isocyanate, 160
Carbenes:
   addition to an alkene, 67
   addition to alkenes, 65
   insertion, 66
Carbocation:
   formation, 28
   relative stability, 99
   stability, 40
Carbohydrates:
   reaction with NaOH, 182
   reactions, 179, 180, 181, 182, 183, 191
   stereochemistry, 183
Carbon-13 NMR, use in unknown identification, 200
Carboxylic acids:
   acidity, 145, 146, 148
   bromination, 134, 199
   esterification, 141
   reaction with diazomethane, 151
   reactions, 138, 153, 155
   reaction of silver salt, 140
   relative acidity, 144
   synthesis, 136, 142, 143, 150, 152, 154, 177, 184
   synthesis from alkenes, 149
Charge, calculation of, 1
Chlorination, alkanes, 22
Chlorohydrin, synthesis from alkenes, 59
Claisen condensation, mechanism, 176
Conformational analysis, 14
   decalin, 19
Conjugated systems:
   cycloadditions, 75
   reactions, 74
   synthesis, 78
Cyanogen bromide, cleavage of peptides, 187
Cyanohydrin formation, 112
Cycloadditions, 75

Cycloalkanes, 14, 21, 43
Cyclohexane, stereochemistry, 14

Decalin, conformational analysis, 19
Decarboxylation, 140, 156
   mechanism, 173
Dehydration, alcohols, 42
Dehydrohalogenation, mechanism, 41
Diastereomers, 183
Diazomethane:
   mechanism of reaction with carboxylic acids, 151
   reaction with carboxylic acids, 151
   use in ring expansion of cyclohexanone, 111
Diazonium salts, decomposition, 165
Dimerization, alkenes, 47
Dinitrofluorobenzene, 186
Diols:
   reactions, 105
   synthesis, 53
Dipole moments, 5

Edman reagent, mechanism of action, 188
Electrocyclic reactions, mechanism, 193, 194, 195, 196
Electrophoresis, 185
Elimination, 35, 36, 73
   mechanism, 41
Enols, 87
   reactions, 175
Epoxides:
   reaction with Grignard reagent, 97
   synthesis from alkynes, 68
Ester, hydrolysis, 136
Esterification, 93
   carboxylic acids, 154
   mechanism, 141
Esters:
   acidity of alpha hydrogens, 156
   chlorination, 22
   hydrolysis, 142
   hydrolysis reactions, 184
   reaction with Grignard reagent, 97
   reactions, 130, 174, 176, 178
   reduction with lithium aluminum hydride, 117
   synthesis, 199
Ether hydrolysis, mechanism, 101
Ethers, 11
   formation, 48
   reaction with sulfuric acid, 8
   synthesis, 100

Favorskii Rearrangement, 133
Fischer projection, 61
Formula:
   determination from percentage composition, 2
   representation, 3
Formula writing, Fischer projection, 61
Friedel-Crafts alkylation, 79, 81
Friedel-Crafts reaction, rearrangement, 85

Gabriel synthesis, 168
Glucaric acid, 183
Grignard reaction, 78
Grignard reagent:
   basic behavior, 87
   mechanism of action with conjugated systems, 118

reaction with carbonyl carbon, 118
reaction with conjugated systems, 118
reactions, 97

HBr, addition to conjugated systems, 75
Hell-Volhard-Zelinsky reaction, 134
Hemiacetals, mechanism of formation, 110
Hinsberg test, 162
Hoffmann elimination, 164
Hofmann reaction, mechanism, 170
Hunsdiecker reaction, mechanism, 140
Hybridization, geometric implications, 4
Hydrazones, reactions, 116
Hydroboration, alkenes, 49
Hydrogenation:
   alkenes, 50
   stereochemistry, 60
Hydrolysis, nitriles, 143
Hydroxyesters, Reformatsky synthesis of, 123

Iodoform reaction, mechanism, 131
Isocyanates, use in Carbaryl synthesis, 160
Isomerism, 6, 10, 11, 13
   dicyclorocyclopentane, 20
Isomerization, aldose-ketose, 182

Jasmone, structure determination, 132

Ketals:
   formation, 117
   hydrolysis, 117
   as protecting groups, 117
Ketones:
   acetal formation, 115
   catalytic reduction, 60
   cyanohydrin formation, 112
   decarboxylation, 156
   oxidation via Baeyer-Villiger reaction, 121
   oxime formation, 113
   reaction with diazomethane, 111
   reaction with Grignard reagent, 97
   reaction with NaOH, 122
   reactions, 117, 119, 130, 175
   reaction with Wittig reagent, 120
   reactivity, 127
   Schiff base derivatives, 114
   synthesis, 119, 157, 178
   synthesis from alcohols, 109
   synthesis from alkynes, 69
   synthesis from diols, 105
Knoevenagel reaction, mechanism, 174
Kolbe reaction, mechanism, 150

Lactones:
   formation, 138
   synthesis, 199
Leaving group, comparison of, 38
Lipids, 184

Malonic ester synthesis, 177
Mechanism:
   acetal formation, 115
   acetate pyrolysis, 197
   addition of HBr to conjugated systems, 84
   alcohol dehydration, 42, 78
   alcohol formation via aldol condensation, 127

alkene bromination, 52
alkyne hydration, 69
amide formation, 171
anion rearrangement, 175
Baeyer-Villiger reaction, 121
base promoted isomerization of carbohydrates, 182
Beckmann rearrangement, 147
benzoin condensation, 126
bromination of alkanes, 23, 25, 26
bromination of alkenes, 197
bromination of propane, 24
Cannizzaro reaction, 129
Carbaryl formation, 160
carbene addition to alkene, 65
carbocation formation, 28, 30, 99
chlorination of alkanes, 22, 25
chlorination with thionyl chloride, 96
chromic acid oxidation of alcohols, 109
Claisen condensation, 176
decarboxylation, 173
diazonium salt decomposition, 165
diol formation, 54
diol rearrangement, 105
electrocyclic reactions, 193, 194, 195, 196
elimination, 35, 36, 41
elimination reaction, 34
esterification, 141
esterification with diazomethane, 151
ether hydrolysis, 101
Favorskii Rearrangement, 133
Friedel-Crafts alkylation, 79
Grignard reagent and conjugated systems, 118
HBr addition to alkenes, 56, 58
Hemiacetal formation, 110
Hofmann reaction, 170
Hunsdiecker reaction, 140
Iodoform reaction, 131
Knoevenagel reaction, 174
Kolbe reaction, 150
ninhydrin reaction, 190
nitration, 82
nitrile hydrolysis, 112, 143
nucleophilic aromatic substitution, 89, 166, 169
nucleophilic displacement, 98, 125
nucleophilic substitution, 95
olefin oxidation by peracid, 106
oxime formation, 113
peptide cleavage via BrCN, 187
peptide cleavage via Edman reagent, 188
phase transfer reactions, 88
phosphorylation of ribose, 191
Pictet-Spengler synthesis, 172
polymerization, 64
rearrangement, 130
reduction of alkenes, 198
Reimer-Teimann reaction, 104
ring contraction of cyclic alpha diketone, 122
ring expansion via diazomethane, 111
RNA hydrolysis, 192
SN1 vs. SN2, 29
SN2 reaction, 31
substitution, 35, 39
substitution reaction, 32, 33
sulfonate ester acetolysis, 139
sulfonate ester formation, 102
Wittig reaction, 120
Wolff-Kishner reduction, 116

INDEX TO PROBLEMS

Methionine, identification in peptide, 187
Methyl isocyanate, synthesis, 158

Ninhydrin reaction, mechanism, 190
Nitration, 82
Nitrile hydrolysis, mechanism, 112, 143
Nomenclature:
    alcohols, 17
    alkanes, 18
    bicyclic compounds, 16
Nucleophilic aromatic substitution, 89
    mechanism, 166
Nucleophilicity, order of, 27
Nucleophilic substitution, internal, 156

Oxidation, via peracids, 106
Oxime formation, mechanism, 113
Oximes, reactions, 147
Ozonolysis, alkenes, 51, 55

Peptides:
    Bergmann synthesis, 189
    identification, 185, 186, 188
    N-terminal identification, 186
Phenols:
    acidity, 108, 145
    reactions, 150
Phosgene, reaction with methylamine, 158
Phosphorylation, mechanism, 191
Pictet-Spengler synthesis, mechanism, 172
Polymerization:
    free radical, 63
    mechanism, 64
Proteins, 185
Pyridine:
    nucleophilic aromatic substitution, 169
    reactivity *vs.* pyrrole, 163
Pyrrole, reactivity *vs.* pyridine, 163

Rearrangement:
    Beckmann, 147
    carbocations, 165
    Favorskii, 133
    Friedel-Crafts reaction, 85
    mechanism, 175
Reformatzky reaction, 123
Reimer-Teimann, mechanism, 104
Resonance, 6, 12
    steric inhibition of, 108
Ribose, phosphorylation by ATP, 191
Ring expansion, Tiffeneau-Demjanov, 157

RNA, basic hydrolysis, 192
RNA hydrolysis mechanism, 192

Sandmeyer reaction, 159
Schiff base, formation, 114
Sedoheptulose:
    reactions, 180
    structure, 180
Sigma complex, nitration, 82
Simmons-Smith reaction, 62
Sodium metal, reactions, 91
Solvolysis, 30
Spectrometric analysis, rearrangement product, 139
Spectrometric data, 132
Spectroscopy, 199
Stereochemistry, 37, 44
    alkene bromination, 52
    bromination of alkenes, 197
    carbene addition to alkenes, 65
    carbohydrates, 183
    chlorination, 96
    chlorohydrin formation, 59
    diol synthesis, 53
    elimination reaction, 34
    epoxide synthesis, 68
    Fischer projection, 61
    hydrogenation, 60
    nucleophilic substitution, 98
    SN2 reaction, 31
    substitution reaction, 32
Substitution, 35, 37, 38
Sucrose, hydrolysis, 181
Sugars:
    reactions, 180
    reducing, 179
    structure identification, 180
Sulfa drugs, synthesis from aniline, 161
Sulfonate esters:
    acetolysis, 139
    synthesis, 102

Tautomerization, keto-enol, 72
Tiffeneau-Demjanov ring expansion, 157
Triglycerides, hydrolysis, 184

Williamson synthesis, 100
    anisole, 103
Wittig reagent, mechanism of action with ketones, 120
Wolff-Kishner reaction, mechanism, 116